Giles David Moss

Pharmaceuticals—
Where's the Brand Logic?
Branding Lessons and Strategy

*Pre-publication
REVIEWS,
COMMENTARIES,
EVALUATIONS . . .*

"So far the pharmaceutical industry has used brand names but not managed products as true brands, nor exploited the profit of a real brand logic. This practice has now reached its limits. Thanks to Giles Moss's very timely book, the directions to adopt and their implications are clear and straightforward. A must-read."

Jean-Noel Kapferer
Professor at HEC in Paris;
Author, *The New Strategic Brand Management*

"A must-read for senior management facing the new challenges in health care as well as the individuals who aspire to positions of product and marketing management. But those who will gain immediate value from this book are the practitioners involved today in promoting and delivering these therapies to the patients and health care providers. After reading this book they should think of themselves more as branding managers, rather than brand/product managers."

Laurence G. Poli, MBA, PhD
Managing Partner,
Center for Performance
Excellence, LLC

More pre-publication
REVIEWS, COMMENTARIES, EVALUATIONS . . .

"At last we have a book that not only focuses on brands and branding from a pharmaceutical perspective but that is both highly readable and highly insightful. Make no mistake about it, this is not simply a rehashing of consumer brand theory for a pharmaceutical audience; on the contrary this is a well-argued and challenging work that takes our understanding of what pharmaceutical branding could achieve to new levels. For me this work should be compulsory reading for all who aspire to market or play a part in the marketing of pharmaceutical brands.

Moss takes us through a journey that begins with the Greek and Roman origins of branding, builds a real understanding of how the theory and practice of branding can be applied and adapted to the pharmaceutical industry, and ends by challenging us not to shy away from tough decisions about branding but to embrace brand theory and take a long-term strategic view in the interests of both the individual product brands, the companies who make them, and the consumers who use them. Whilst the first five chapters provide an excellent analysis of how current theory and practice apply to the pharmaceutical industry it is the second half of the book that really excited me, as it fundamentally questions our current approach to pharmaceutical branding; Moss confronts the assumption that the pharma brand can only survive as long as its patent and articulates the benefits of putting the pharma brand at the heart of our enterprise, removing the rigid focus on the product life cycle and opening our minds to the potential to sustain a brand over a much longer period of time.

His arguments here are compelling and we ignore them at our peril; hopefully the debate on rebuilding the 'tarnished pharmaceutical industry brand' begins here."

Mike Owen, MMRS, BSc
CEO Brand Health International

"Giles has created a book based on his deep understanding of the pharmaceutical industry, to which he has applied a marketer's eye. He has found an industry that has focused on product features, with revenue driven largely by a heavy emphasis on sales promotion."

Gary Noon, MSc, BSc, MBA
CEO, Aegate Limited,
Cambridge Technology
Centre, Melbourn

"A fascinating take on pharmaceutical branding from an industry insider. This book expertly recites the popular models of consumer branding, drawing on such savants as Aaker. It adds to the body of knowledge and will be a popular addition to the shelves of numerous pharmaceutical company executives."

Tom Blackett
Group Deputy Chairman,
Interbrand; Co-editor,
Brand Medicine: The Role of Branding in the Pharmaceutical Industry

More pre-publication
REVIEWS, COMMENTARIES, EVALUATIONS . . .

"**E**xcellence and rigorous process are essential in today's pharmaceutical environment. *Pharmaceuticals— Where's the Brand Logic?* is essential reading for executives seeking a practical toolkit for brand management and provides insightful approaches to extracting the maximum value from products in the face of cost-containment and regulatory, public, and media scrutiny. This book supports our reflection on the evolution of the pharmaceutical industry and its future brand strategies."

Leonard Lerer, MD, MBA
INSEAD Healthcare
Management Initiative;
Managing Editor,
Journal of Medical Marketing

Pharmaceutical Products Press®
An Imprint of The Haworth Press, Inc.
New York

NOTES FOR PROFESSIONAL LIBRARIANS
AND LIBRARY USERS

This is an original book title published by Pharmaceutical Products Press®, an imprint of The Haworth Press, Inc. Unless otherwise noted in specific chapters with attribution, materials in this book have not been previously published elsewhere in any format or language.

CONSERVATION AND PRESERVATION NOTES

All books published by The Haworth Press, Inc., and its imprints are printed on certified pH neutral, acid-free book grade paper. This paper meets the minimum requirements of American National Standard for Information Sciences-Permanence of Paper for Printed Material, ANSI Z39.48-1984.

DIGITAL OBJECT IDENTIFIER (DOI) LINKING

The Haworth Press is participating in reference linking for elements of our original books. (For more information on reference linking initiatives, please consult the CrossRef Web site at www.crossref.org.) When citing an element of this book such as a chapter, include the element's Digital Object Identifier (DOI) as the last item of the reference. A Digital Object Identifier is a persistent, authoritative, and unique identifier that a publisher assigns to each element of a book. Because of its persistence, DOIs will enable The Haworth Press and other publishers to link to the element referenced, and the link will not break over time. This will be a great resource in scholarly research.

Pharmaceuticals—
Where's the Brand Logic?
Branding Lessons
and Strategy

Pharmaceuticals—
Where's the Brand Logic?
Branding Lessons and Strategy

Giles David Moss

Pharmaceutical Products Press®
An Imprint of The Haworth Press, Inc.
New York

For more information on this book or to order, visit
http://www.haworthpress.com/store/product.asp?sku=5836

or call 1-800-HAWORTH (800-429-6784) in the United States and Canada
or (607) 722-5857 outside the United States and Canada
or contact orders@HaworthPress.com

Published by

Pharmaceutical Products Press, an imprint of The Haworth Press, Inc., 10 Alice Street, Binghamton, NY 13904-1580.

PUBLISHER'S NOTE
The development, preparation, and publication of this work has been undertaken with great care. However, the Publisher, employees, editors, and agents of The Haworth Press are not responsible for any errors contained herein or for consequences that may ensue from use of materials or information contained in this work. The Haworth Press is committed to the dissemination of ideas and information according to the highest standards of intellectual freedom and the free exchange of ideas. Statements made and opinions expressed in this publication do not necessarily reflect the views of the Publisher, Directors, management, or staff of The Haworth Press, Inc., or an endorsement by them.

Cover design by Kerry E. Mack.

Library of Congress Cataloging-in-Publication Data

Moss, Giles David.
    Pharmaceuticals—where's the brand logic? : branding lessons and strategies / Giles David Moss.
        p. cm.
    Includes bibliographical references and index.
    ISBN: 978-0-7890-3258-4 (hard : alk. paper)
    ISBN: 978-0-7890-3259-1 (soft : alk. paper)
    1. Drugs—Marketing. 2. Brand name products—Management. 3. Pharmaceutical industry. I. Title.
HD9665.5.M66 2007
615.1068'8—dc22
                                                                                        2006036619

For Joanne, Annabel, and Sebastian.

ABOUT THE AUTHOR

Giles D. Moss, MBA, MRPharmS, BSc, is a pharmaceutical industry insider who has risen through the ranks during a twenty-year career. He worked his way up from Sales Representative at Squibb, his experience encompassing virtually all positions in marketing and sales management (BMS, Sandoz, SmithKline Beecham) before moving on to general management. He works in Operations as Vice President Europe, Region 1 for UCB S.A., a global top five biopharmaceutical company. Previous recent experience includes Regional General Manager South East Asia, Australia, and New Zealand and Head of CNS Global Marketing and Medical Affairs, both at UCB. He has published pharmaceutical brand articles in numerous publications including the *Journal of Brand Management,* the *International Journal of Medical Marketing,* and the *Journal of Pharmaceutical Marketing & Management.*

CONTENTS

Foreword

"Food for thought" is a commonly used phrase, but what is often served up is neither tasty nor intellectually nourishing; not so with what Giles Moss has served up. Of late, there have been several books dedicated to the idea of branding in pharmaceutical markets; most are offered up by advertising executives anxious to bring their consumer product magic to the world of pharma. However, these books tend to be filled with examples of mega-consumer brands with admonitions of the pharmaceutical industry for failing to achieve the long-term success of the Tides, Crests, Marlboroughs, and the like— as if patent expiration was merely a date on the calendar rather than a fundamental market shift.

I have spent a significant amount of time and energy in recent years focusing on the differences between the markets for consumer products and those for prescription drugs, and have reached the con- clusions that the differences, while not insurmountable, are sufficient to prevent the direct application of most consumer marketing ap- proaches to pharma with any real success. Still, there is much to be learned from all aspects of marketing and the theories on brand and franchise management. Its focus on portfolios of business and lon- gevity should be understood by all who engage in pharma marketing.

I have been struck by the recurring observation that many of the people involved in pharmaceutical marketing lack a fundamental un- derstanding of the theories and concepts of marketing, and display an ignorance of the body of literature upon which they are built. Phar- maceutical marketing, in practice, is more a trade than a profession, where the practitioners learn from "journeyman" marketers, who learned from those who came before them. Rather than building a marketing plan based on a foundation of marketing theory, most plans are simply a set of activities that have been undertaken by every

Pharmaceuticals—Where's the Brand Logic?
Published by The Haworth Press, Inc., 2007. All rights reserved.
doi:10.1300/5836_a

previous product team, with the addition of the latest technology or gizmo to make it a little more exciting. The lack of application of fundamental marketing understanding to most pharmaceutical products, it seems, has served the practitioners well, although I believe many products—and firms—have suffered. But when the rules of the game are to "do unto products what others have done before," one would be brave, or foolish, to do otherwise. John Maynard Keynes, the much-maligned economist of the first half of the twentieth century, observed, "Worldly wisdom teaches that it is better for reputation to fail conventionally then to succeed unconventionally." This, I believe, is alive and well in the pharmaceutical industry. Fortunately, a few practitioners, such as Giles Moss, believe and behave otherwise—and live to tell their stories!

I first met Giles Moss several years ago when he invited me to develop and deliver a seminar on life-cycle management to a diverse global group at his company. He was very specific about his needs and concerns, and made no bones about his disappointment in the "generic" presentations by many business faculty—he needed something his people could use. Through our discussions leading up to the presentation, it was clear that I was dealing not just with a practitioner, but a student of pharmaceutical marketing with a collection of case studies and an inquiring mind. Months later, when he had taken on new responsibilities, he asked me about the potential for a book on pharmaceutical branding and brand management, and I urged him to go for it—and I am thrilled with the result.

E. M. (Mick) Kolassa

Preface

Why write about pharmaceutical brands? The consumer world in which we live has changed dramatically over the last twenty years; brands have more significance and impact in every part of our daily work and home life. International brands are seen everywhere around us, branding is discussed on television, airport bookstores offer strategic branding books that are best sellers—branding has become a phenomenon. Strategic branding theory has progressed, and its implementation has driven growth in consumer consumption around the world.

In contrast, branding within the pharmaceutical world hasn't progressed at the same pace. Branding is not treated as a science or a discipline, and is often thought of as something best reserved for advertising agencies. Marketing people in the industry now talk about brands and branding but haven't grasped the significance of major changes seen in fast moving consumer goods (FMCG). The relevance of branding is still difficult to evaluate in the pharmaceutical industry, which is driven by large research and development (R&D) budgets, vast sales forces, and patent protection lawyers. At this moment, the industry is under more pressure than ever before to maintain sales growth, and at the same time is under constant attack from activists and governments critical of how we conduct our sales and marketing activities. Brand building is very far from the minds of our industry thought leaders.

This book charts five years of personal discovery on the topic of brands and is written from the context of someone who has spent more than twenty years and all his working life in the pharmaceutical industry. Therefore, it illustrates how a pharmaceutical sales and marketing man has tried to fathom out what branding means in the

Pharmaceuticals—Where's the Brand Logic?
Published by The Haworth Press, Inc., 2007. All rights reserved.
doi:10.1300/5836_b

FMCG world and then related it to the world he knows. When I first started researching brand theory and asked "Who are the real experts?," it was easy to identify the consumer brand gurus, but when I asked around within the pharmaceutical world, the most frequent reply was only one or two names, or more commonly that "all the experts are practitioners within the industry." The purpose of this book is to reopen the debate around the significance that strategic brand building could attain in pharmaceuticals in the future. The text has been guided by nonpharmaceutical thinking and hopefully will provide practical help to those struggling with the issues we all come across on a day-to-day basis. The real richness of experience that has guided my conclusions has been difficult to put down on paper—especially as many of my personal experiences and opinions would probably put me in court!

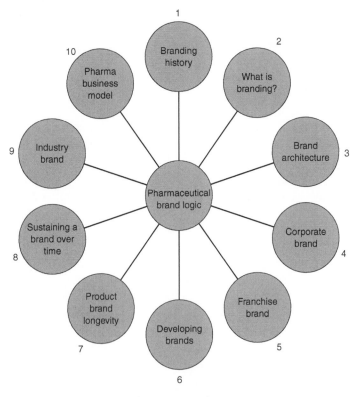

Chapter overview.

The first chapters of the book are devoted to the practicalities of branding, starting with the history of branding in Chapter 1, explaining how the pharmaceutical business doesn't easily fit into the various consumer branding models that exist. Chapter 2 looks at what branding is and how classic brand management has evolved over the decades into a more strategic discipline, a significant step forward from the largely tactical origins that the pharmaceutical industry still maintains. Chapter 3 sets the scene for the real meat of the book, looking at brand hierarchy and starting to set a framework around what branding could and should mean to the pharmaceutical industry. Chapters 4 and 5 cover the corporate brand and the franchise brand before a more individual product brand perspective is taken in the Chapters 6, 7, and 8, covering, among other subjects, global branding, product brand longevity, and sustaining a brand over time.

Chapters 9 and 10 of the book take a look at the wider issues within which brands have to provide value and relevance within the pharmaceutical industry. Chapter 9 covers the pharmaceutical industry brand, its decline at the end of the twentieth century, and what has to be done to try and resurrect it before it becomes too late. The final chapter looks at the business model the pharmaceutical industry employs and some of the challenges it faces.

If you agree with the content, that's great, but if you disagree that's even better, as we need to raise the bar collectively to benefit as an industry. Please give me feedback at www.pharmabrandlogic.com.

Chapter 1

Branding History

The origins of branding can be traced back over 2,000 years, to the times of the Greeks and Romans. Initially in those days, little product identification was required, as most goods were made locally and the purchaser had a close relationship with the supplier of the goods. As the Roman Empire expanded, the distance of the product's point of sale from its source increased; this led for the need for "marks" to establish the origin of a product. They acted as an indicator of quality from the individuals that created them and, later, along with signatures when added to bowls or manuscripts, they formed the first brand marks and became the starting point of where we are today.

This situation was thought to be relatively stable until the industrial revolution of the eighteenth and nineteenth centuries, which heralded the manufacture of mass-produced items. Many aspects of what we see today as branding were created during this time. Advertising to the general public helped create many famous brands that thrive to this day (e.g., Cadbury's, Coca-Cola, Colgate, Gillette, Lipton's, Heinz, and Quaker Oats). The new brands grew in success, and, over the following decades, as the original founders of the business grew older and retired, responsibility for protection of the brand moved from individuals (who in many ways still provided a personal guarantee, like thousands of years earlier) to large corporations and multinationals. The brands themselves therefore became the focus, and brand management was born in the fast moving consumer goods (FMCG) sector, where most of the marketing and advertising innovations happened.

Pharmaceuticals—Where's the Brand Logic?
Published by The Haworth Press, Inc., 2007. All rights reserved.
doi:10.1300/5836_01

Pharmaceuticals, to some extent, moved in parallel during this period. A number of founders started their pharmaceutical businesses in the late eighteenth century, such as Thomas Beecham with his famous Beechams Powders preparation. But, by the time the major innovations of the twentieth century were made, many of the founders had retired, the business had grown through acquisitions, and the responsibility had passed to large organizations and multinationals such as SmithKline Beecham. A second example would be Sandoz, which was founded in 1886 by Dr. A. Kern and E. Sandoz as a chemical company, during the same decade that saw Coca-Cola invented by a pharmacist in Atlanta. Pharmaceutical substance production started in 1895, and later major milestones included the 1982 introduction of Sandimmun (an immunosuppressant) and the merger with Ciba Geigy in 1996 to form Novartis. The Sandoz brand then disappeared for a short spell before being reborn in 2003, uniting all the generics businesses of Novartis under one single brand name.[1]

The 1980s and 1990s saw a proliferation of FMCG brands in the consumer market place, as communication and transportation between markets became more achievable and affordable. The business and the consumer became even more separated by distance (and now time zones), and this coincided with changes within the communist bloc, which allowed multinationals to market their consumer products globally for the first time. But these new consumers, rather than merely adopting established western brands, and therefore generating new sales, were bombarded by local brands and local own-label brands, which evolved with surprising speed. There was an increase in the competitiveness of each market, making what should have been a simple geographical expansion, more difficult.

Proliferation of brands was also seen in the established consumer goods markets in the West, first from the original manufacturers and then followed by the major retailers launching distributor own brands (DOBs). These distributor own brands started as cheaper, lower-quality copies of the original brands, increased the potential for confusion in the minds of the consumer. Consumer goods brand owners clung to the hope that as people became confused by the choice of DOBs (generics), established original brands would guide them through. Recognition of the importance of branding occurred with these battles against the generics, and the concept that "the brand is everything" was born. Financial recognition of the value of brands also took place, being recorded as financial assets within the accounts of leading brand owners. The significant sums paid for luxury goods brands in the 1990s gave the final stamp of approval for the discipline.

Branding as a discipline has started to mature and a significant amount of research and thought has gone into the subject during the last thirty years. The brand work that has been completed relatively recently shows that managing brands should essentially be seen as a tool by which to develop the business and its resultant profitability. The differences between retail (FMCG) brands, service brands, and business-to-business brands are starting to be investigated, and a number of innovative companies have challenged the norm and boundaries of brand building.

FMCG brands have become increasingly powerful and are well suited to these global brand-building activities, as they are relatively easy to control, due to limited interactions between the customer and the total brand experience. They have had to fight back against distributor own brands that have risen in quality, from the low-quality offerings of the 1980s, and which now represent a significant threat to the originals they imitate. The new millennium brought with it an increasing awareness of the availability of fake goods, sourced in particular from Asian countries such as China and South Korea, and this now represents the next battleground for the FMCG sector.

Service brands are recognized as being more difficult to control, as they have to rely on the branding of a process, and the people-based delivery of the service is less predictable and affected by inconsistency. A typical service brand could be represented by an airline (or hotel, shopping center, etc.) where the combination of say the price, seat, in-flight entertainment, and flight attendant service make up the brand offering. As a result, the complexity of getting it right increases, but the ability to differentiate various offerings from other airline service providers also intensifies. Other service brands would be Internet service providers such as Google, Yahoo!, and Amazon, which have grown into huge businesses in a relatively short period of time through innovative branding to suit their individual business models. Their brands are based on attitudes rather than specific products; you are attracted to the Web site and then might be sold virtually anything, ranging from books to DVDs and cameras.

Business-to-business (B2B) brands are normally characterized by high-tech and industrial brands and are seen as the most complex branding area.[2] Product purchase decisions can be highly complex, with a significant degree of commercial risk attached. These products are often bought by people who are not the ultimate users of the product or service, and must appeal to a very wide range of influencers at different times during the purchasing process. The acquisition of expensive capital equipment such as a mainframe computer can often be led by a technical specification list that is generated by coordination across multiple departments. The equipment then has to fit into a certain capital budget and subsequent training and operational expenses budgets for the future. The final decider may be someone who never uses or maintains the item in question and merely hands over the organization of contracts, delivery, and even payment terms to other departments after the purchase decision has been made. There is, therefore, some similarity to branding within pharmaceuticals in that there are multiple influencers, and the complexity of the final decision is high, as illustrated in Figure 1.1.

These three classical categories for brand management, FMCG, service, and B2B don't easily fit in with current thinking in pharma-

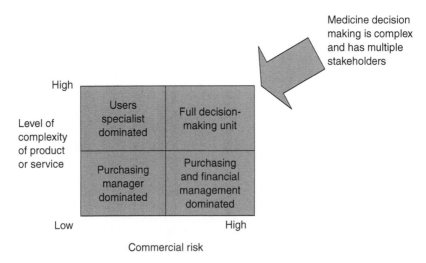

FIGURE 1.1. Business to business (B2B) brand model. (*Source:* Adapted from Ellwood, in turn adapted from de Chernatony and Macdonald.)

ceutical branding, and this in itself is the source of much industry confusion. Pharmaceutical marketers have to skillfully integrate different parts of the classical categories to achieve their results. Direct-to-Consumer advertising (DTCa) and Internet communication are close to FMCG practice, but are much more regulated. Many pharmaceutical activities are service based (such as provision of clinical outcomes data and disease management advice), whereas the traditional customer of the industry (the physician) works on a branding model closer to B2B than anything else. The prescribing decision is highly complex, the product is used by a third party, and there are many other influencers on the decision making process nowadays, from formularies, insurance institutions, or reimbursement decisions to key opinion leader advice. As a result of this complexity and the need to demonstrate efficacy, safety, and tolerability to achieve marketing authorizations, we have fallen into the product attribute trap of marketing. Figure 1.2 illustrates some of the complexity seen in the prescribing decision in the twenty-first century.

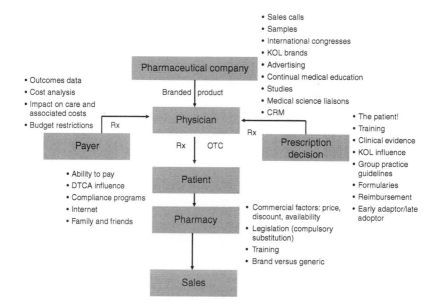

FIGURE 1.2. The pharmaceutical buying process.

THE PHARMACEUTICAL PRODUCT ATTRIBUTE TRAP

Classical high-tech, industrial, and pharmaceutical marketing has assumed that customers are only interested in the technical attributes a product has, and base their decisions solely on selection of those attributes. As a result, pharmaceutical marketing probably has closer links to these areas than the more exciting and easily identifiable FMCG sector—although, in time, with the advent of DTCa for medicines, this will change. The problem for pharmaceutical branding, and what it means to the industry is that, so far, out-of-the-box thinking hasn't really been seen to be successful.

Looking at a closely related sector, a good example of innovative brand building is the high-tech brand Intel—a brand that has cleverly raised the issue of what's inside the PC box, as a way to differentiate a basic semiconductor commodity. It also gained control over microprocessor standards at the same time. The "Intel inside" logo and tune are known worldwide and provide a reassurance for non-technical people who have to make complex purchase decisions.

The early marketing of a pharmaceutical tends to focus quite strictly on the basis of the product license: its indications, how the product is administered, and the established efficacy, safety, and tolerability, which have been proven through registration studies. If the product has additional noteworthy characteristics, like being first in a new class due to having a different mode of action, then this would most likely also be featured in early promotion. A simple, nonexhaustive illustration is shown below for the lipid-lowering brand Zocor (simvastatin), which was originally launched at the end of the 1980s as the second "HMG" product brand launched in the United States market and the first within Europe. At that time, cholesterol-lowering agents still suffered from the stigma of early clinical trials that potentially linked increased rates of suicide with cholesterol reduction.

- First-in-class HMG-CoA reductase inhibitor (in Europe)
- Licensed to reduce LDL cholesterol and increase HDL cholesterol
- New mode of action

- Reduces LDL cholesterol by 18-32 percent using 10-40 mg
- Tolerability and safety so far measured as equivalent to placebo
- 10-20 mg tablet once daily

Until recently, most product brands had to generate their early sales from registration data. The funds to invest in significant additional clinical trials in advance of first commercialization were difficult to find, and further development of the brand then occurred through additional studies. These would deepen the published experience with the product brand, broaden the indications of its license, and develop the dosage regimen with a view to strengthening the competitive claims in the market. The major change for Zocor came more than five years after launch, when the 4S study allowed a whole raft of new claims, including, most significantly, positive health outcomes end points, to strengthen its positioning, for example:

- Reduces the risk of total mortality by reducing coronary events
- Reduces the risk of nonfatal myocardial infarction
- Reduces the risk of stroke or transient ischemic attack

The investment made by MSD to generate this type of outcomes data was huge and took a long time to arrive, but it gave the added benefit to Zocor, and to the whole class, of revolutionizing how payers around the world viewed the importance of cholesterol reduction. Indirectly, and in concert with other major outcomes studies with other "HMG" product brands such as Lipostat (pravastatin), from BMS, the results paved the way for explosive growth. Pfizer's Lipitor, which arrived at the end of the 1990s, rapidly took market share and achieved sales in excess of $10 billion in 2004.

PRODUCTS, NOT BRANDS

Classical marketing, as previously illustrated, consists of managing products, not product brands. But, as we know from consumer research, brands have an existence beyond the product, and, therefore, consumer brand management has undergone significant change over the last two decades. It follows other principles and concepts and uses other tools than classical marketing. Consumer brand marketing shows that the emotional proposition is vital. We know that, as consumers, even when the brand fails us from time to time, we can forgive if the emotional attachment is strong enough. Our loyalty isn't just a sum of the rational reasons for brand selection. Figure 1.3[3] illustrates some of the differences between a pharmaceutical product (illustrated by product attributes) and a pharmaceutical brand.

One of the few examples of published physician research carried out by Kapferer[4] concludes that a product gives a certain efficacy, whereas a brand gives trust; a brand has a certain personality that gives additional values to the product versus the competition. Kapferer also looked at the relationship between what doctors are exposed to (as far as clinical papers and advertising are concerned), what their other influencers are (such as key opinion leader effect and the age of the product), and then came up with the relationship shown in Figure 1.4. The shaded and bold entries have been added to try and include some of the newer aspects involved in medicine marketing.

His research suggested that brand personality and quality image of the product were the key factors that created a status for the drug.

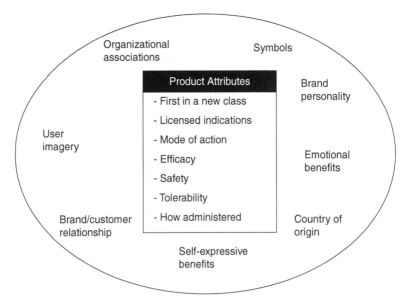

FIGURE 1.3. Brand.

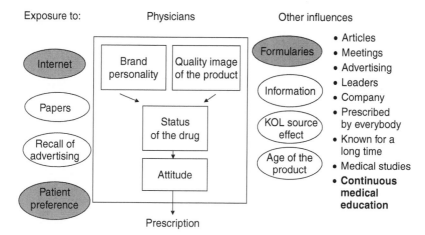

FIGURE 1.4. Physician exposure and their influencers. (*Source:* Adapted from Kapferer, 1997.)

This then led to an attitude to prescribe and eventually the use of the brand for a patient. This early work by a nonpharmaceutical specialist identified the major influences on a physician in the 1990s. Since then, the situation has been further complicated by patient involvement in decision making, the impact of information available on the Internet, more stringent formularies, and significantly increased continuous medical education (CME).

The work of Owen and Chandler has further developed the pharmaceutical understanding of brands. The large number of medicines available to the modern physician to treat diseases means that brands play an important role in differentiation. This is particularly true when the product attributes are very closely matched.[5] In a disease area such as hypertension, a primary care physician (PCP) could choose from among 100 different therapies within eight drug classes. The differences between these medicines may be miniscule within a class and minor between the classes; therefore, brands are needed to take a more significant role in choice. Building a clearly communicated brand will provide differential advantage in crowded therapy areas because the prescriber needs help for their decision making.

The last two decades have seen an increasing understanding of how the brain works. Developing the work of neuroscientific researchers,

Owen and Chandler have explored the relationship between mind and culture to create "neural maps" of how physicians think about pharmaceutical product brands.[6] Figure 1.5 sets out some of their thinking on the subject, garnered from the research.

Their work concludes that what a pharmaceutical product brand means to a physician is created from an interaction between their memories and emotions, as well as their interaction with the medical culture and its rules and meanings. Brands therefore become central to understanding the product. This also leads to the conclusion that emotion can be a powerful part of the decision making process for physicians, especially when it is making up for knowledge gaps in complex situations.

In the past, some prescription drug marketers have believed that giving a name to a certain product will make it a brand. Others remark that many practitioners believe merely adding a bit of symbolism to a product will be sufficient to create a brand. Despite this simplistic approach, pharmaceutical products do become brands, even if their development is often not planned: The patient will, through experience, gain emotional benefits, whereas the prescribing physician will gain both emotional and self-expressive benefits (e.g., I use the best medicines for my patients) as well as developing other associations. Some pharmaceutical product brands will develop rich brand personalities (e.g., oncology or antidepressants), where the impact on life expectancy or quality of life is large, whereas others where the usage is more mundane will not (e.g., athletes foot ointments).

Neuro-scientific understanding	Cultural context
• Most mental activity is not fully conscious • Most memory is not immediately accessible or conscious • Invisible "stored" memory plays a key role in what we do • Emotion and reason link to guide what we do	• People require common ground to live together • Culture is a mixture of rules and meanings that guide us • Everyday-life reinforces those rules and meanings, but they are also adjusted and bent

What a brand means to an individual is created from memories, emotions, rules, and meanings that create a system of understanding

FIGURE 1.5. The neuro-cultural view of brands. Brands become central to understanding the product.

Put simply, patients (and doctors) will have significant interest in their medicine, because they want to get better (or cure patients).

Traditionally, the pharmaceutical industry has focused on products and not necessarily on product brands. The logic behind this concerns a relatively short patent life (and/or narrow exclusivity window), and a rapidly changing medical environment, which makes it essential to move onto the next product offering from the R&D pipeline. The industry at the end of the 1990s talked only about blockbuster products, how to make them bigger, and how to achieve blockbuster status faster. The discussion was not about establishing brands or the different aspects of managing brands to lead the industry into the uncertain future.

As companies have merged, blockbusters have had to get bigger just to deliver historical rates of sales growth. In 1998, Tagamet was the world's best seller with $1 billion sales; in 1995, Zantac achieved $3 billion sales; in 2000, it was Losec with $5 billion; and Lipitor exceeded $10 billion in 2004. In general, blockbusters are global products, although examples of sole U.S. blockbusters do exist, and this global commercialization has been the most effective tactic over the last two decades.

Figure 1.6 illustrates how speed of uptake has accelerated at the same time as size of the top blockbusters has increased.

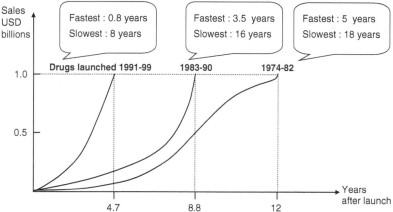

FIGURE 1.6. Speed to reach $1 billion sales has accelerated. Analysis of 45 blockbusters launched between 1974 and 1999. First country launched taken as starting date. (*Source:* Adapted from oral presentation of F. Mascha [McKinsey] "Key factors to make a pharmaceutical product a mega brand.")

The perceived reason for this urgency is that pharmaceutical products are thought to be relatively short-lived in comparison to industrial, service or consumer goods brands (in fact this is not necessarily true as you can see in the chapter on sustaining a brand over time). Pharmaceutical product life cycles are measured in years rather than decades or indeed centuries. Figure 1.7 illustrates the longevity of some consumer brands within the United States. Sixty-four percent of the most well-known brands in the United States are fifty years old or more, with 10 percent being older than 100 years.

A popular view is that pharmaceutical products, in comparison, are limited to fifteen to twenty years exclusivity within the major world markets and, by the time they have been launched, many of those years have already been used up. In addition, it is commonplace to think that after expiry they are rarely actively promoted once generic competition has arrived. (This is another topic that will be discussed in much more detail in the chapter on sustaining a brand over time.) Major exceptions to the rule do exist, some due to continued regulatory exclusivity (versus patent protection), or others due to a lack of approvability of competitors. No real mechanism appears at present to approve generic versions of the large-molecule biotech product brands launched during the 1980s and 1990s.

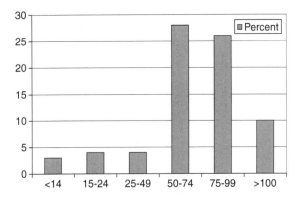

FIGURE 1.7. Longevity FMCG of brands. The age of the most well-known brands in the United States. (*Source:* Adapted from Aaker, D.A. (1991) *Managing Brand Equity,* The Free Press. In turn adapted from Bogart and Lehman "What makes a brand name familiar.")

Looking at an alternative measure of the lasting value of consumer brands would involve category leaders and their relative positions over time; e.g., in America, nineteen out of twenty-two categories had the same leading brand in 1985 as in 1925 (see Table 1.1).

By comparison, within pharmaceuticals, no product has survived with comparatively the same sales value intact. Aspirin, acetaminophen (paracetamol), and codeine preparations may have retained product volume around the world, but none have managed to maintain the same levels of profitability as when they were within patent. Building pharmaceutical brands therefore appears, on the surface, not to protect long-term sales and profit in the same way as it does for other industry sectors.

In general, the pharmaceutical industry has come relatively late to branding. During the 1980s and 1990s, the pharmaceutical industry enjoyed success over an extended period of time, achieving relatively easy double-digit growth on a consistent basis. By and large, this was through using traditional methods, and there was no apparent urgency to change the way it marketed its products. The success of the industry relied on three factors: strong research and development (R&D), aggressive defense of patents, and use of the dominant promotional tool—powerful sales forces. The industry has therefore been driven by product and R&D, and not by market or brand. De-

TABLE 1.1. U.S. brand leaders in 1925 and 1985.

Category	Leader in 1925	Position in 1985
Biscuits	Nabisco	Leader
Cereals	Kellogg	Leader
Cameras	Kodak	Leader
Canned fruit	Del Monte	Leader
Sewing machines	Singer	Leader
Soap	Ivory	Leader
Soft drinks	Coca-Cola	Leader
Toothpaste	Colgate	Leader
Tea	Lipton	Second
Razors	Gillette	Leader

Source: Adapted from the Committee on the value of advertising, American Association of Advertising Agencies (1989) "The value side or productivity: A key to competitive survival in the 1990s," p.18.

spite the size of the sales generated, there were sixty-seven block-busters or products that generated in excess of $1 billion in 2003; drugs have been largely treated as products and not as brands.

The easy growth environment has, however, changed. Industry growth has been slowing down, and firms have been searching for ways to maintain it. The three traditional success factors of the industry are less evident than in the past. First, it has become much more difficult to identify the blockbuster drugs that can fuel company momentum (additionally, product innovation remains costly and more illusive than ever). Second, many of the most successful drugs will soon suffer patent expiry. More than half of the global top fifty best sellers will go off patent in the next five years. Moreover, in view of the concentration of sales in fewer big products, the sales at stake are much larger than in the past. Third, sales force effort is reaching a saturation level as the industry consolidates. It will not be possible in the future to base success just on increasing the number of sales representatives promoting a product.[7]

Combined with this backdrop, generic competition has also been developing rapidly and constitutes an increasingly real threat for the industry. Generic companies benefit not only from patent expiration, but also from the cost reduction pressures evident in every health care system around the world.

The industry has not realized that it is managing brands and not just products. It's clear that a pharmaceutical product has all the elements that make it a brand. It represents in consumers' minds a set of tangible and intangible benefits. It does not only deliver a certain efficacy (tangible) but also offers additional values such as trust (intangible). The brand has an existence in both doctors' and patients' minds, which goes beyond the product itself. Pharmaceutical companies develop molecules, but doctors prescribe product brands.[8]

Chandler and Owen suggest that there is a hierarchy of brand functions with pharmaceutical product brands (see Figure 1.8).[9] Their research has identified that a product brand provides authentication and differentiation at its simplest level. Above that level, it delivers need fulfillment and the establishment of a contract with the physician. At the most complex height of the hierarchy, a product brand also provides orientation and develops charisma to further strengthen its impact.

Viagra research example

- Authentication—it is Viagra from Pfizer
- Differentiation—it is Viagra, not a herbal or injectable

- Need fulfillment—Viagra works
- Contract—patients and their partners having their relationship helped

- Orientation—Allowing impotence to be discussed
- Charisma—a straightforward solution to a difficult problem

FIGURE 1.8. Brand function hierarchy.

Therefore, the erectile dysfunction brand Viagra provides authentication that it is the original brand from Pfizer and the differentiation that it is not a herbal remedy or a more invasive injectable. At the next level, it provides need fulfillment for the physician due to its efficacy and the contract it establishes is with patients and their partners, having helped their relationship. Through the publicity surrounding its launch, the brand allowed changes in orientation, as talking about impotence became acceptable, which then allowed it to develop charisma as a straightforward solution to an extremely difficult problem. This type of research starts to unlock the potential advantages that could be accrued by making branding more scientific and a true discipline in the future.

The industry has reacted to a more difficult environment via consolidation. In a series of significant mergers and acquisitions, it has attempted to maximize R&D and reach economies of scale in the sales and marketing area. Despite this, it is unlikely that mergers will be sufficient in themselves to allow a return to the double-digit growth seen during the 1990s. One of the aspects that will be discussed in this book is that the pharmaceutical industry accepts brand destruction within its business model; it's not at all unusual to see the withdrawal of virtually all support for a pharmaceutical product brand as patent expiry approaches, irrespective of the variation in the brand's life cycle in different markets around the world.

Branding, however, represents a new competitive advantage that could be leveraged by the industry, in line with the success seen in the FMCG area. A great deal has been learnt about pharmaceutical product brands over the last decade, as demonstrated by the work of Owen and Chandler. Branding is not just an advertising agency activity but should be a strategic discipline practiced at all levels. Branding strategies could then help maximize return on investment for new product brands, while helping to offset the inevitable growth of generics in the future. The rest of this book looks at the various aspects of how to develop a "brand logic" for pharmaceuticals.

CONCLUSIONS

Branding can trace its roots back thousands of years to the Greek and Roman empires. The modern brand was born at the end of the eighteenth and beginning of the nineteenth centuries. Significant changes to consumer markets took place in the 1980s and 1990s with the introduction of distributor own brands and the globalization of trade. This lead to adoption of brands as key strategic assets within a company's portfolio, and kick-started the development of branding theory to appreciate FMCG, service, and B2B brand models. Pharmaceutical branding doesn't easily fit into any of the models due to its inherent complexity.

Classical high-tech, industrial, and pharmaceutical marketing has assumed that customers are only interested in the technical attributes a product has, and base their decisions solely on selection of those attributes. Pharmaceutical products do become product brands, but their development is often not planned, whereas, in the past, the industry has focused on creating global blockbuster products to drive double-digit annual sales growth.

In comparison to consumer brands, which heavily use emotional arguments in their communication, pharmaceutical brands appear to be limited by short exclusivity periods and have suffered from total withdrawal of investment (brand destruction) once a generic version becomes available. Pharmaceutical branding theory has advanced over the last decade, creating a clearer understanding of the role of the pharmaceutical product brand, helped by advances in neuroscientific thinking. We know there is a hierarchy of brand functions for physicians that range from simple authentication and differentiation, through need fulfillment and contract forming to orientation and creation of charisma.

The pharmaceutical industry has so far been successful with classical marketing techniques allied to an R&D focus, aggressive defense of patents, and overuse of the key promotional tool—large sales forces. The environment is now changing, and merging to maximize R&D productivity and achieve economies of scale in the sales and marketing area isn't going to be the sole long-term solution for the industry.

By developing strategic pharmaceutical brand logic, pharmaceutical branding could move from being an advertising agency activity to become a tool to develop the business and safeguard future profitability, i.e., a strategic discipline. Branding could provide a competitive advantage by maximizing return on investment for new product brands while offsetting the inevitable growth of generics in the future.

Chapter 2

What Is Branding?

Branding can be extremely confusing to the pharmaceutical marketer. What all the consumer books talk about isn't easy to relate to day-to-day activities. The most useful article on the subject appeared in the *Journal of Brand Management* and was written by a member of the editorial board, Dannielle Blumenthal.[1] In just a few pages, she beautifully describes the reasons for the confusion around a brand and, therefore, branding—it is the direct result of a turf war between a whole host of interested parties who want to make money out of the concept. There is no universal agreement on what a brand is, what you can expect to get from it, or how to make it more profitable for your company. It is in the interests of brand gurus, brand consultants, management consultants, organizational development consultants, advertising agencies, and business schools to be seen as the experts in this highly visible area. Everyone is busy selling their particular niche expertise, claiming unusual insight into a complicated area, presumably making themselves piles of cash, but muddying the waters as they go.

Having said that, it's worth stating a few definitions as a way of illustrating that there are definitions of real value out there and they can make sense. They aren't quite picked at random:

- Branding is "a set of consistent processes, aimed at a specific purpose, that define, differentiate, and add value to the organization."[2]
- A brand "is a name or a symbol given to a product that will differentiate it versus other products and that will register it in the

mind of consumers as a set of tangible (rational) and intangible (irrational) benefits."[3]

- A brand "is the aura that surrounds a product or service that communicates its benefits and differentiates it from the competition for the consumer."[4]
- Brand management is a process that requires resources and focus from all functions, within a company, to succeed in one strategic intent—creating a difference. The fundamental task is to transform the product category.[5]
- "Brand equity is a set of assets (and liabilities) linked to a brand's name and symbol that adds to (or subtracts from) the value provided by a product or service to a firm and/or that firm's customers." The major asset categories being brand name awareness, brand loyalty, perceived quality, and brand associations."[6]

Given that there is general confusion concerning branding, even in the fast moving consumer goods (FMCG) area, there is little real surprise that an industry dominated by mass sales forces, multiple customers, patent defense lawyers, and huge R&D organizations is going to find branding difficult to pin down. The brand concept, as indicated in the above definitions, goes beyond the product concept. A product delivers certain tangible benefits. A brand offers additional values that are both tangible and intangible, adding emotion to rational choice. A brand can be considered as the added value of marketing investments made in a certain product.

The Mercedes brand, for example, is registered in the minds of consumers as a brand offering a set of tangible benefits (solidity, reliability, and Germanic quality) and intangible benefits (success, status), linked to what the brand represents. Some brands focus more on tangible benefits (Volvo, Gillette, and Pampers) while others more on intangible benefits (Coca-Cola, Perrier, and Marlboro) dependant on their market. All have managed to create strong associations in the consumer's mind.

Two notions are of particular importance to understand the brand concept: the brand identity and the brand image. The brand identity is the set of tangible and intangible benefits that the firm has to select to differentiate its product versus competitive ones. The identity is therefore based on a company decision. The brand image is the per-

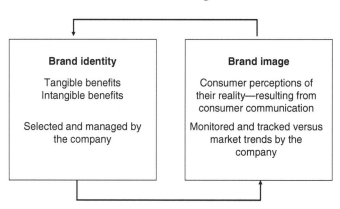

FIGURE 2.1. Brand identity versus brand image.

ception of this identity in the minds of consumers. It is based on consumer perception and is illustrated in Figure 2.1.

WHAT IS A STRONG BRAND?

One of the most influential brand gurus is David A. Aaker, the author of numerous articles and books and a leader in global branding thinking. In his book *Building Strong Brands,*[7] Aaker enlarges upon a concept already quoted earlier in this chapter—brand equity. "Brand equity is a set of assets (and liabilities) linked to a brand's name and symbol that adds to (or subtracts from) the value provided by a product or service to a firm and/or that firm's customers." The major asset categories are:

- Brand name awareness
- Perceived quality
- Brand loyalty
- Brand associations

Strong brands in the consumer world possess these different aspects of brand equity and, therefore, provide value both to the customer and the firm that creates the brand. The consumer can gain confidence in the purchase and increased satisfaction with its performance after purchase if the brand delivers more on top of the simple product features. The firm that creates those additional assets gains by efficiencies of their marketing programs, by perhaps improved prices and margins as well as the ability to negotiate strongly with the trade and lever the brand name to allow the successful launch of extensions.

Interestingly enough, within the model of brand equity proposed by Aaker are also "other proprietary brand assets," which are very briefly explained as other sources of customer loyalty or perhaps patents—an area with obvious importance for pharmaceuticals (and something we'll come back to, especially in the chapter on sustaining a brand over time). What is missing perhaps is the emotion behind buying behavior. Consumers have to believe the money spent for a brand is worth it, in the same way a physician has to believe treating long-term illness with chronic medications is worthwhile.

Brand Name Awareness

Looking briefly at the different components of the model and starting with brand awareness, Aaker expands to say that this provides a signal that it is a brand to be considered, it signals substance and commitment from the organization, provides a familiarity for potential liking of the brand, and gives an anchor to which other associations can be attached. All of these components and what they might bring are relatively familiar to pharmaceutical marketers, although what associations we might want to anchor to the brand are often forgotten as we have to focus on the inevitable efficacy, safety, and tolerability messages that form the platform of pharmaceutical marketing. In fact, if there is one thing we are good at, it is creating awareness! We are probably world class at screaming the brand name at the target customer via advertisements, mailings, mass sales forces, and numerous other tactical media, but this is only part of the process.

Perceived Quality

The next component of perceived quality is a more difficult one for pharmaceutical marketers to grapple with. Perceived quality in the consumer world drives a number of critical decisions and can be the reason to buy in the first place; it is often the source of how the brand is differentiated and at what price it can be realistically sold. In addition, it impacts on the ability to launch extensions and whether or not retailers would be interested in stocking the brand in the first place. Quality perspectives do exist within pharma but not in the same context. After all, any prescription medicine has to be granted a marketing authorization to be commercially available. Therefore, the final quality stamp is provided by a third-party group, such as the United States Food and Drug Administration (FDA), which restricts when and how a medicine can be used. In theory, therefore, quality in itself cannot be a driver for purchase of a pharmaceutical, but in reality every physician and pharmacist (and even patients taking chronic medication) know that quality is a factor in usage—quality of the clinical data that backs up a product, quality of the response a company will give if a pharmacovigilance issue arises, quality of their response if a product is deliberately tampered with in the supply chain, and, of course, quality of bioequivalence and release characteristics. Although the industry doesn't discuss quality a great deal, and doesn't use it promotionally as a strong lever, potential quality issues with generics, the threat of counterfeit medicines, and the general inability to maintain continual supply, all become relevant to the stakeholders who decide which pharmaceutical brands should be chosen. This aspect of quality is difficult to convey for individual products, but is probably underused in corporate franchise campaigns. Awareness of "quality" is going to grow in the future, especially as China is tipped to become the single biggest manufacturer of pharmaceuticals within this decade and since organized crime is now known to be interested in the margins available from counterfeiting pharmaceuticals.

Brand Loyalty

Brand loyalty is the next component, and, in consumer terms, it provides reduced marketing costs (retaining a customer is presumably less expensive than having to find a new one), improved trade leverage with the big-name retailers (the competition for space can be intense), and providing awareness and reassurance for new customers as they already know the brand in advance of their first purchase. Brand loyalty is also accepted to buy the firm time to react to new competitive pressures that arrive in the market, as destabilizing the satisfied consumer takes longer than taking a dissatisfied purchaser looking to change brands anyway.

From the pharmaceutical perspective, most of these points have some validity—even improved trade leverage is valid within an over-the-counter (OTC) context, but much less so in prescription pharmaceuticals where the distribution channels provide a delivery service rather than taking a gatekeeper role. Retaining a happy customer (e.g., retaining a prescribing physician via customer relationship management (CRM) is going to be cheaper than converting another, and brand familiarity does provide reassurance for the first time a physician tries a new drug. Brand loyalty can also be said to be driven by the nature and complexity of prescribing medicines, in particular when a patient may be on a variety of medications at the same time. Therefore, if a patient is stable and any resultant side effects have been successfully managed by the physician then there is inevitably a reluctance to change medications on both parts, especially if the disease area is serious and/or unstable or life threatening. Patient loyalty schemes exist in many guises in different markets around the world, whether that is coupons to reduce the purchase price on repeat scripts or compliance or lifestyle advice sent on a regular basis. The consumer principle is the same for physicians—it's easier and cheaper to retain a satisfied customer than to have to convert a new one.

Brand Associations

In the pharmaceutical world, brand associations may have something in common for the different customers within the model or, they may be divergent. Emotional associations of quality associated with a particular product may be shared by the patient, physician, and pharmacist in areas where regular and consistent bioavailability is critical (theophylline treatment for instance). But they may be quite different at other times. A physician may get self-expressive positive personal associations from treating patients with the most up-to-date therapies possible, whereas a pharmacist may get the same personal satisfaction from providing a product with the highest personal profit margin! Reminders of other possible associations you may want to connect to your brand can be provided by the games many pharma PMs will have played in the past at their advertising agencies. If you had to imagine your product as a person, symbol, organization, etc., what would it be?

The pharmaceutical industry spends a huge amount of intellectual and commercial effort in the attempt to create barriers to competition via obtaining patents, periods of marketing exclusivity, or data protection. In this way, they slow or prevent the launch of generic copies of the original brands (whether NCEs or other branded generics). This area in fact characterizes the industry and its attitude that creation of the product is more important than creation of the brand.

Strong brands in the pharmaceutical world provide many of the advantages that strong brands in the consumer world do (see Figure 2.2). They reassure both patient and physician that the right choice has been made. There are few more complex purchase decisions around than treating a human being, who is different from every other human being, with an active pharmacological agent that is known to affect different patient subtypes in different ways. Any confidence provided in the decision-making process is invaluable, in particular as all medical practitioners know from experience that a patient's attitude toward his or her medication is a key ingredient to success. Strong brands deliver in the backup they provide to the medical community, whether that is in the response from a medical information enquiry or in the fact that their supply chain is maintained at all times. Patients gain from numerous compliance programs and advice in trying to achieve their treatment goals and in so doing become more empowered about looking after their own health, rather than being dependant on others

Customer value	Company value
• Reduced risk in drug usage choice for the physician • Increased patient commitment to continued therapy • Increased post-prescription satisfaction for the physician (I've done the right thing–used the best brand)	• Reduced marketing costs (retaining customer cheaper than finding new) • Improved price–possible in numerous markets • Longer-term revenue stream post expiry

FIGURE 2.2. Strong pharmaceutical brands. Brand equity value creation.

for their care. In theory, marketing costs are also reduced, as satisfied customers are less likely to change medications, especially given the complexity of decision-making, and this allows a brand to build loyalty over time. In some markets, strong brands also allow the achievement of a price premium over the competition (free price markets), and, in other price controlled markets, they allow the brand to become part of the equation of the formulary setters of government institutions. In reality, the only areas where strong pharmaceutical brands don't mimic their consumer equivalents is in negotiation with the trade distributors or providing the ability to regularly launch extensions of the brand (something confined to the OTC sector).

MARKETS THAT CHANGE

Brands have to operate within a changing market environment and changing societal culture. Consumers change how they feel about a brand, and marketers change how they position and communicate around a brand. This is also true within pharmaceuticals where change is a desired outcome, due to short periods of exclusivity for pharmaceutical product brands. Chandler and Owen[8] identify four major factors of change within pharmaceutical markets, as shown in Figure 2.3.

The competitor set is the market most easily understood, it is about competing within the old rules of the market, identifying the strengths and weaknesses of each brand, and delivering improvements against those competitors. Market dynamics concern trends inside or outside of a market that can cause an impact. A trend may be the move toward empowered patients taking more decisions about their own health care, or government cost-containment measures slowing the adoption of new product brands into a specific country. Market parameters are seen as major factors for change where they allow the mixing of previous choice criteria with new criteria, thereby changing the overall market. The launch of Imigran (sumatriptan) for migraine, along with the weight of prelaunch education around the severity of the disease, persuaded many PCPs to take the treatment of migraine more seriously. The final factor is probably the most interesting, as it covers market contradictions. These factors are difficult to understand but can be very powerful; they are tensions or points of stress that ad-

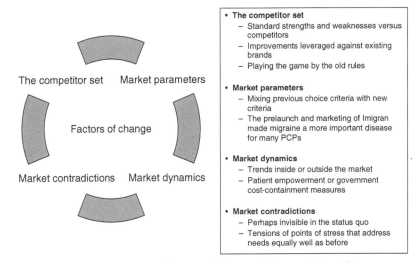

The competitor set   Market parameters

Factors of change

Market contradictions   Market dynamics

- **The competitor set**
  - Standard strengths and weaknesses versus competitors
  - Improvements leveraged against existing brands
  - Playing the game by the old rules

- **Market parameters**
  - Mixing previous choice criteria with new criteria
  - The prelaunch and marketing of Imigran made migraine a more important disease for many PCPs

- **Market dynamics**
  - Trends inside or outside the market
  - Patient empowerment or government cost-containment measures

- **Market contradictions**
  - Perhaps invisible in the status quo
  - Tensions of points of stress that address needs equally well as before

FIGURE 2.3. Factors of change within markets.

dress needs equally well as before. They might initially be invisible in the status quo but once activated can have a dramatic effect. Lipitor (atorvastatin) came to market without long-term outcomes data that showed reduced risk of mortality. Until that point, the cholesterol-lowering market had been driven by outcomes availability, and the competitive set assumed that, without such proof, conservative physicians around the world would be slow to adopt the new product brand. The contradiction in the market that allowed Lipitor to rapidly become the world's biggest selling product brand was the "know your number" factor within the patient and physician populations. Lipitor's greater efficacy meant that long-term outcomes data, which in reality was only a hope in the mind of the physician, was overshadowed by dramatic short-term reductions in cholesterol numbers.

CLASSIC BRAND MANAGEMENT

Branding folklore suggests that classic brand management, as far as the consumer goods industry is concerned, started at Procter & Gamble (P&G) in the 1930s. In order to try and solve "sales problems," P&G decided to use brand management teams that were charged with the responsibility of marketing a particular product and coordinating activity across departments, mainly sales and manufacturing. In those days, this meant conducting research, finding out the cause of poor sales, and then developing programs in response to improve the situation. Each product was allowed to compete with the other products within the company, and the role was therefore largely tactical, the brand team reacting to competitor moves with new campaigns and margins. The system allowed many peoples' activities to be guided, and kept everyone moving in one direction. It is the system by Aaker and Joachimstaler,[9] outlined in Table 2.1, which has been seen to be both successful and widely imitated.

According to Kapferer,[10] brand management is a process that requires resources and focus from all functions within a company to succeed in one strategic intent—creating a difference. The fundamental task is to transform the product category.

FMCG firms dedicate a lot of management attention and efforts to manage their brands. Branding is a strategic priority at every level of the organization. Brands are created as part of the research and development process, and marketing people work in depth with R&D right

TABLE 2.1. Pharmaceutical brand logic—the evolving paradigm: The classic pharma brand management model.

Perspective	Tactical and reactive
Brand manager status	Less experienced, shorter time horizon
Conceptual model	Brand image
Focus	Short-term financials
R&D	Late consultation
Product-market scope	Single products and markets
Brand structures	Simple
Number of brands	Focus on single brands
Country scope	Single country
Brand manager's communication role	Coordinator of limited options
Communication focus	External/customer
Driver of strategy	Sales and share

Source: Adapted from Aaker and Joachimstaler (2000).

from the very beginning of the product development process. This teamwork is aided by the fact that R&D is a relatively inexpensive and fast process, normally measured in months or perhaps a few years. The focus can therefore be on creating brands that will last decades, and the marketing of consumer brands can maximize the long-term brand growth.

The previous summary is not that distant from modern practice in the pharmaceutical industry. Pharmaceutical marketing spends a huge amount of its time looking at "sales problems" and coming up with programs to solve those problems. Changing "the category" equates in our industry to trying to change the competitive claims set within a narrow therapeutic class of medicines (e.g., the proton pump inhibitors [PPIs], the lipid lowerers, the antiepileptic drugs [AEDs], etc.). In that way, the product can take the upper hand and, therefore, gain market share more rapidly, allowing higher sales before generic competition arrives. Few pharmaceutical products are managed within a complementary portfolio and, therefore, are expected to compete versus other company products—in fact, often as an act of cannibalization at the end of a products patent. Within the antiepileptic class (category) of products, Pfizer has the first generation Dilantin (phenytoin), the second generation Neurontin (gabapentin), and has recently launched Lyrica (pregabalin) in the hope it will cannibalize the sales of Neurontin before generic competition takes too much of its share. In reality, therefore, there are many similarities between classic brand management and modern day practice within the pharmaceutical industry.

However, there are a number of fundamental differences in the ways the industries operate: (1) the size of the sales forces involved, (2) the complexity of the modern-day pharmaceutical company, and (3) the extended time involved in interacting successfully with R&D. The time required to gain competitive advantage for a marketed product, via the generation of outcomes data, for instance, or in the even more extremes cases the preparation of a new product launch is a lengthy process. The focus, therefore, tends to be the product, rather than the brand, due to the amount of resources and time it takes to create it in the first place. This is compounded by the relatively short period available before patent expiry, to maximize the sales return rather than build a long-term brand.

MODERN BRAND MANAGEMENT

Modern brand management argues that the classic brand model falls short in dealing with emerging market complexities, competitive pressures, global forces, multiple brand portfolios, and complex sub-brand structures. The classic brand management model was created a long time ago, and globalization forces over the last twenty to thirty years have aged it considerably. The new Aaker imperative termed "Brand Leadership" emphasizes strategy and vision rather than merely tactics and reactivity, and as a result tries to cope with some of the complexities of modern day business.

Table 2.2 shows some of the differences between the classical brand management model and the new brand leadership model. Aaker and Joachimstaler believe that brand management needs to be handled in organizations by more senior individuals who are able to hold a longer-term horizon, and that brand equity measures should replace short-term financials in a way that allows the role to be strategic (rather than merely tactical). They also see the need to move from a limited focus to a broad focus, dealing with multiple products and markets, complex brand architectures, and issues driven by globalization. The brand manager needs to become a team leader with multiple communication options (both internal and external) to utilize. Finally, they believe the driver of strategy should be brand identity, rather than sales and market share.

This newer model starts to address some of the major issues we face, issues that stop consumer branding models being useful within the pharmaceutical environment. It is clear, the industry is in the middle of a strategy crisis, falling NCEs, harsher market conditions, and changing consumer expectations, to name but a few which we are failing to deal with. If branding is going to have an impact on this industry, it will have to be led from the top, and at this stage senior industry managers are often not the most talented marketers within their organizations. The pharmaceutical industry has to learn to cope with being a global business, extracting value from its vast merged portfolios and defining clear roles for multiple products in the same category (or therapeutic area). The move from sales and market share to brand identity being the driver of strategy is probably way too far for the industry at this stage, but it doesn't mean that we couldn't adopt some of those measures for our softer corporate, or franchise

TABLE 2.2. Pharmaceutical brand logic—the evolving paradigm.

	The classic pharma brand management model	The brand leadership model (FMCG)
From tactical to strategic management		
Perspective	Tactical and reactive	Strategic and visionary
Brand manager status	Less experienced, shorter time horizon	Higher in the organization, longer time horizon
Conceptual model	Brand image	Brand equity
Focus	Short-term financials	Brand equity measures
R&D	Late consultation	
From a limited to broad focus		
Product-market scope	Single products and markets	Multiple products and markets
Brand structures	Simple	Complex brand architectures
Number of brands	Focus on single brands	Category focus—multiple brands
Country scope	Single country	Global perspective
Brand manager's communication role	Coordinator of limited options	Team leader of multiple communication options
Communication focus	External/customer	Internal as well as external
From sales to brand identity as driver of strategy		
Driver of strategy	Sales and share	Brand Identity

Source: Adapted from Aaker and Joachimstaler (2000).

brands. Table 2.3 makes some suggestions about what brand management could look like in the pharmaceutical industry in the future—it adapts the more strategic approach (we know the tactical traditional approach has its limitations) and raises the importance attached to it within the bigger organization.

Therefore, pharmaceutical brand management would become more strategic and practiced at a more senior level. It would need to be suitably analytical to cope with the complexity of portfolio management (including the R&D interaction) rather than maintaining an individual product brand focus.

CAREER PATHS IN MARKETING

The traditional career path to reach general management in FMCG is to grow in the marketing function to first become a brand manager, then a category manager, and finally a marketing director. The FMCG

TABLE 2.3. Pharmaceutical brand logic—the evolving paradigm of the future.

	The classic pharma brand management model	The brand leadership model (FMCG)	The pharma-ceutical brand logic model
From tactical to strategic management			*Strategic management*
Perspective	Tactical and reactive	Strategic and visionary	Strategic and visionary
Brand manager status	Less experienced, shorter time horizon	Higher in the organization, longer time horizon	Higher in the organization, longer time horizon
Conceptual model	Brand image	Brand equity	Pharma brand equity
Focus	Short-term financials	Brand equity measures	Short/medium-term financials & brand equity for corporate & franchise brands
R&D	Late consultation		Early consultation
From a limited to broad focus			
Product-market scope	Single products and markets	Multiple products and markets	Multiple products and markets
Brand structures	Simple	Complex brand architectures	Complex brand architectures
Number of brands	Focus on single brands	Category focus—multiple brands	Corporate/franchise and product brands
Country scope	Single country	Global perspective	Global perspective
Brand manager's communication role	Coordinator of limited options	Team leader of multiple communication options	Team leader of multiple communica-tion options
Communication focus	External/customer	Internal as well as external	Internal as well as external
From sales to brand identity as driver of strategy			
Driver of strategy	Sales and share	Brand identity	Sales/share and brand identity

Source: Adapted from Aaker and Joachimstaler (2000).

marketing function is considered as a line job but is sited in the center of the organization, unlike the pharmaceutical industry, where global marketing is a staff function, and the sheer size of the sales forces in each affiliate means marketing support is required in the countries. As a result, country marketing receives the majority of resourcing in pharma, leaving a gap at the center of the organization.

In the pharmaceutical industry, the organization of brand management is quite different to that seen in the consumer world. Global marketing people will often come late into the development process, often in phase 3b, close to final registration. It's clear, however, that key decisions are taken at a much earlier phase of the product's development plan, often years earlier when the product enters phase 2. The primary endpoints for registration studies are defined by regulatory guidance requested from agencies such as the FDA and as such are a necessity for commercialization. On the other hand, many of the secondary endpoints within these studies can, with careful thought, have a powerful impact on the claims a product can be launched with. Crafting physician as well as payer stories can be done at an early stage, and this realization has started to change attitudes in some of the bigger companies such as AstraZeneca, GlaxoSmithKline, and Lilly. But this doesn't really constitute brand management within a strategic context, although it is getting closer to the practices of the consumer goods industry.

Pharmaceutical marketers are often more sales-driven than marketing-driven and therefore pay more attention to the executional elements of marketing rather than developing the strategic thinking that is required to make in-depth analyses of data from the market, the consumers, and the competitors. The traditional marketing career route to the top in the industry is to start as a representative, then follow this with country-specific product management (sometimes for as little as a couple of years), and probably then back to sales in a management position to allow a career path in the direction of being a country general manager. The ability to manage vast numbers of people within a sales function is critical to career progression, as powerful sales forces so far have been the dominant promotional tool within the industry. Operational top management therefore has tended to come from individuals who have experienced big line management careers rather than a specialized marketing background. If you then add to this top-tier senior R&D management, who have only worked in that area and have the necessary finance and legal expertise, and the lonely marketing director, this then constitutes the make up of many operating committees. As a result, marketing experts are on the periphery at the top level, especially as central or global marketing positions often do not hold the same cachet of their counterparts in FMCG and can be relegated from key organizational groups. This traditional career path will make the move to a strategic pharmaceutical brand management model more difficult.

CONCLUSIONS

There is no universal agreement on what constitutes a brand, what branding means as a discipline, or what brand equity provides to an organization, due to a "turf-war" between interested parties, all wanting to be the experts. What brand management should provide is the creation of a difference.

Brand identity is a company-driven communication of the associated tangible and intangible benefits, whereas brand image is what the consumer perceives. Brand equity is a set of assets (and liabilities) that include brand name awareness, perceived quality, brand loyalty, and (importantly for pharmaceuticals) brand associations.

Brands have to operate within a changing market environment and changing societal culture. Consumers change how they feel about a brand, and marketers change how they position and communicate their brand. Pharmaceutical product brand work identifies four major factors of change within pharmaceutical markets. Those factors are the traditional competitor set, market parameters, market dynamics, and market contradictions.

Modern brand management has evolved from the classic tactical P&G model in the 1930s to a more modern version that is more strategic and visionary as it starts to deal with more complex brand architectures and a global perspective.

Strong brands in the pharmaceutical world could provide many of the same advantages of strong brands in the consumer world. They could create brand equity value for both the customer (physician and patient) and the company. Therefore, pharmaceuticals brand management would become more strategic and practiced at a more senior level. It would need to be suitably analytical to cope with the complexity of portfolio management (including the R&D interaction) rather than maintaining an individual product brand focus. Traditional sales and marketing pharmaceutical careers present a potential block to moving to a more strategic brand management model.

Chapter 3

Brand Architecture

INTRODUCTION

In the 1970s and 1980s, branding was relatively simple. A single product from a single company would deliver a single promise, for example, the Bic razor, Coca-Cola (before light or diet), and the Sony Walkman. As a result, most internal and external stakeholders could understand the complexities involved. The brand signals the origins of the product and distinguishes it from other similar product offerings from other sources. But as soon as these successful companies launch a second product, decisions have to be made. Should the new product be associated with the original product brand, with the company that produced it, or would it be better if it were seen as a totally separate brand?

During the 1980s and 1990s, the problem really started to accelerate as multinational companies grew in size, and, as a result, so did the number of brands they marketed. Soon, brands developed into complex portfolio structures of interrelated product brands, which if not managed in a coherent fashion could easily confuse customers and employees of how they fitted in. Mergers and acquisitions accelerated the process, and, as a result, proliferation of brands was the major issue in the 1990s. Sony was represented by the Walkman twenty years ago, and now this brand successfully stretches across multiple categories from TVs, to music, electronic games, movies, and even insurance and banking in its home Japanese market.

The two major contributors in the discussion about how to deal with portfolios of brands have been Jean-Noel Kapferer and David

Pharmaceuticals—Where's the Brand Logic?
Published by The Haworth Press, Inc., 2007. All rights reserved.
doi:10.1300/5836_03

Aaker. The story started in 1992, when Kapferer first published his classic brand book titled *Strategic Brand Management,*[1] in which he discussed six levels of branding, using the term "brand architecture." He talked about product brands, line brands, range brands, umbrella brands, source brands, and the endorsing brands. In a series of books starting in 1996, Aaker contributed three terms to this topic, starting with "brand systems," moving to "brand architecture" (like Kapferer) before finally settling on "brand portfolio strategy" in his latest 2004 offering.[2] Aaker, for instance, identified different driver roles for brands in portfolios, such as strategic, endorser, and the silver bullet, and then analyzed how these brands might be leveraged.

The following illustrates the essence of the complex decision-making process surrounding successful portfolio management of each brand:

> ... [Brand architecture] specifies the structure of the brand port-folio and the scope, roles, and interrelationships of the portfolio brands. The goals are to create synergy, leverage, and clarity in the portfolio and relevant, differentiated, and energized brands. (Aaker)[3]

This topic is probably most relevant to the pharmaceutical indus-try—many of the reasons why "branding" isn't taken seriously in pharmaceuticals is that we haven't developed an understanding that all of our separate independent brands have a scope, a role, and an interrelationship in the wider portfolio. Just as the consumer goods world has seen proliferation of brands over the last two decades, so has pharmaceuticals—perhaps for very different reasons—but the impact is the same. Virtually all the top-fifty pharmaceutical compa-nies have highly complex portfolios consisting of hundreds of sepa-rate pharmaceutical products, some of which have become true brands (through design or accident). In those portfolios are vast numbers of "off patent" products whose sales history is measured in decades. Many of these "tail" products provide stable sales from protected niches or chronic treatment areas. Still more are used by only small segments of the physician population and will eventually die out when that particular age group of physicians finally retires. The immense multibillion-dollar portfolios that now characterize the industry have been built up through years of mergers and acquisitions; companies that started out as prescription-only pure players have added OTC divisions, animal health divisions, medical devices divisions, and even divisions selling genetically modified food.

A number of "mixed" model companies still exist, despite their efforts to simplify their structures. This type is most obviously illustrated by Bayer, which still maintains chemical, pharmaceutical, and crop science divisions. Another more successful mixed player, UCB S.A., has recently completed its transformation by divesting its films and chemicals divisions while acquiring the premier British Biotech company Celltech to become the fifth largest biopharmaceutical company globally. These kinds of bold moves are however not common (excluding mergers), and the management of complex portfolios across the industry appears at this time to be at best rudimentary and at worst purely a financial process (being pushed and pulled by the predominant opinions in the financial community at any particular juncture). Whereas a huge amount of intellectual effort goes into driving the sales of a few blockbuster product brands and acquiring new products through business development, little real thought is evident about the portfolios in which they play a part. Using a biology analogy, a pharmaceutical company's brands should fit together forming a single organism (portfolio), in which each cell (brand) performs a specific function and role, thereby allowing the larger organism (company) to thrive.

The traditional pharmaceutical hierarchy of branding, which is trotted out at innumerable pharma industry conferences, is that of product, franchise, and corporate branding. The most significant area of activity to date is, without doubt, the product branding segment.

As discussed in previous chapters, the pharmaceutical industry has tended to focus on technical functional attributes and has managed to bypass much of the consumer world's increasing sophistication, until recently. We largely see pharmaceuticals as a collection of separate product brands but occasionally refer to individual companies in a franchise fashion (FMCG equivalent is a category). In that way, MSD is a cardiovascular house, BMS an oncology house, and the old Smith-Kline Beecham a vaccines house. Corporate brand building in the industry has been relatively limited. When these activities have been undertaken, they have largely been reserved for the signaling of new merger names (e.g., Aventis and Novartis, or some basic attempts at image rebuilding, GlaxoSmithKline corporate adverts).

What appears to be becoming clearer and clearer is that an industry brand also exists, the brand of the pharmaceutical industry itself (i.e., the image and perceptions held by consumers [patients], health care professionals, and even the investment community of the wider industry). The problem is that the industry brand image is worsening over time. Considering the amount of good this industry has done over the past 100 years, the situation we find ourselves in is horrible. Chapter 9 looks at why the pharmaceutical industry perceived any better than the tobacco and oil (gasoline) industries. One of the reasons for this dire position is perhaps that the industry focus on product branding has been to the detriment of other branding strategies, such as building strong corporate brands that are capable of endorsing product brands. The unmanaged industry brand has therefore been easy to attack, and weak corporate brands do not have the strength of depth to maintain some facet of goodwill under concerted attack from activists.

Figure 3.1 represents how many in the industry currently perceive branding priorities, with the industry portion at the top (left), overestimated in the visual. Product branding therefore gets the vast majority of resources, but is seen as an advertising agency activity, whereas the industry or corporate brands receive no focus. The right-hand side

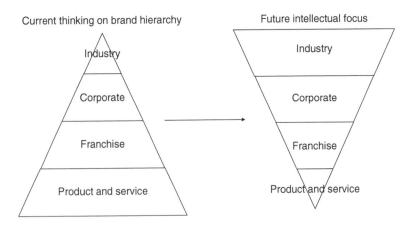

FIGURE 3.1. Pharmaceutical brand hierarchy.

of the figure shows where the challenge is in the future. It isn't just a matter of spending some millions in advertising budget; instead, it's about making the intellectual effort to make a difference. The industry needs to understand that investment in rebuilding the industry brand and establishing strong corporate brands are prerequisites for successful franchise and product brands. This may be an intellectual leap of faith and a seismic shift for an unsophisticated branding audience, but it's a vital component for protecting the future.

To try and add some light to this difficult subject, this chapter will look at the work of Kapferer and Aaker as far as brand architecture is concerned. It will look at whether their models might fit current pharmaceutical working practice and provide some insights for the future. The chapter will cover a review of the architecture of branding, look at management of pharmaceutical portfolios and look at how they can be leveraged to the advantage of the business.

BRAND HIERARCHY

As mentioned previously, Kapferer identified six main models for branding strategy based largely as a function of two aspects of brand building, i.e., the brand identifying the source of the product and the brand contributing to product differentiation (see Figure 3.2). Kapferer argued that significant thought has to be given to the structure of the overall portfolio so that it aided buyer recognition and understanding by being logical. In this way, it would also guide the larger originator organization to make the right decisions through the use of rules concerning naming, symbols, colors, etc. The figure adapted from Kapferer shows the six strategies.[4]

Product Brand

The advantages of product brand strategy are that it is an offensive way of occupying a whole market segment; e.g., the number of Procter & Gamble (P&G) detergent brands that are seen to be independent from each other includes Tide, Cheer, Bold, Ariel, and Dash. First-mover advantage accrues to the first product in a new category, and the brand can thus become the absolute reference for the category in question (e.g., Hoover). In addition, as a result of the company re-

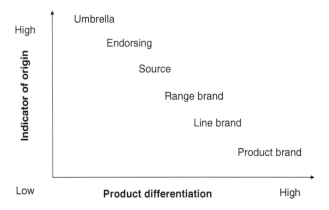

FIGURE 3.2. FMCG brand strategy hierarchy. (*Source:* Adapted from Kapferer, Jean-Noël [1991], *Strategic Brand Management,* pp. 189, The Free Press, New York.)

maining largely anonymous in using this strategy, a company may feel it can take more risks. Similarly it could perhaps protect in times of crisis as it allows free movement for the company across categories with less chance of related damage if a product withdrawal is necessary. The disadvantages of such a strategy are that any association with a successful renowned brand and any possible economies of scale in terms of sales and marketing are difficult to achieve.

Not surprisingly, the product brand appears to be very close to the pharmaceutical model. Over an extended period of often decades, this strategy has been used to dominate certain therapy classes (e.g., ICI's and Zeneca's presence in hypertension, angina, and heart failure). By legal requirement, each new medicine (new chemical entity [NCE]) has its own unique name. In fact, it has two, one brand name and one nonproprietary (or generic) name. In most cases, the pharmaceutical product brand is largely independent, with very few having corporate name associations, and, as a result, they remain anonymous, except where legal identification of origin is required on the packaging and label. The argument that this may allow more risk taking does not really stack up however. Whereas the eventual patient may not be aware of the origins of a specific product brand, the prescribing physician and associated medical professions often are. No one seriously thinks that, worldwide, those in the know don't associate the withdrawals of Lipobay and Vioxx with Bayer and MSD. Pharmaceutical companies are not constrained by their categories (or franchises) as new products are launched into any disease area that appears to be worthwhile, and this is emphasized further by the largest pharmaceutical company worldwide only having a market share of approximately 10 percent (way below that of the car makers, for instance). Economies of scale are, therefore, more difficult to achieve in pharmaceuticals, due to the nature of very fragmented portfolios (but this is one of the main advantages of the franchise brand, which will be discussed later). But some economies of scale can be accrued in pharmaceutical sales and marketing, particularly as the key promotional tool is sales force one-to-one selling, and each sales person can promote a number of product brands at the same time.

Brand Extensions

A brand extension is defined in branding theory as an existing brand name being extended to a category of products that is different from the existing one. A line extension consists, on the other hand, of the launch of new products, under the same brand name, in the same product category.

In FMCG, the use of brand extension has been frequent and has been increasing over the last ten years. In view of the very high cost of launching new brands and managing them, firms have decided to launch new products behind existing brand names. This builds on the trend to concentrate efforts on big brands only. For example, Procter and Gamble is concentrating on big brands that generate more than $1 billion sales. They have recently decided to launch two new innovations under existing brand names. New biodegradable wipes, named Kandoo under the Pampers "umbrella" name and a new product for washing cars under the Mr Propre/Clean "umbrella" name. This trend can be seen in both multinationals and local companies.

Line Brands

The consumer idea behind this strategy is that if you have a successful product and distinctive positioning, it is possible to offer many complementary products under the single name. Therefore, it is achievable to exploit the concept by successfully extending in the form of a line brand and gain from a stronger brand image, easier distribution, and lower launch costs of the new associated line. The line can be launched as a brand or built up over the years but to be true to its definition it needs to be close to the original product, i.e., in its product class. Done well, line brands can expand the user base, inject variety and innovation, and block competitor activity (e.g., L'Oreal's Studio Line of hair care products). Done badly, it can create costs, reduce focus, and cause confusion.

In the pharmaceutical world, the very fact that every new chemical entity has its own name limits the possibilities for line brands to be largely just different formulations of the same product brand. The Canestan line has a number of separate formulations widely available to consumers such as pessaries, tablets, and creams. They allow the patient to use the same active ingredient in different ways to achieve the same end. Another example is illustrated by carbamazepine, a first- generation antiepileptic agent. It is available as immediate release tablets, slow release tablets, chewable/melt tablets as well as a liquid, and as such each formulation appeals to a different subsegment of its target patients. Figure 3.3 compares FMCG and pharmaceutical line brand extensions.

Regulatory agencies (such as the FDA and the European Agency for the Evaluation of Medicinal products [EMEA]) don't allow the retention of a brand name for a next generation product, and this reduces the efficacy of the line brand concept in the pharmaceutical world (e.g., Prozac 2, Lipitor Max, or Viagra Plus would not be approved as NCEs). As a result, transferring and leveraging the brand heritage of one product brand for another has to be done in less obvious

• Line brands FMCG	• Line brands pharmaceuticals
– Complementary products launched under a single brand name	– Galenical forms
	– OTC switches
– Line extensions need to be within the product class e.g., hair care for L'Oreal	– Different formulations of the same chemical entity that may include tablets, powders, liquids, slow or immediate release, and OTC forms

FIGURE 3.3. Line brand extensions.

ways. The next generation Nexium (esomeprazole) from AstraZeneca retained a number of design elements including the color of the tablets when it took over from Losec (omeprazole), which had been the biggest selling worldwide product brand in the late 1990s.

The line brand avenue of thought can however be extended to include switches of prescription-only products to over-the-counter (OTC) status, allowing extended exclusivity protection in some markets such as the United States, for example, Pharmacia's Rogaine/Regaine (minoxidil), Schering-Plough's Clarityn (loratidine), or Merck's Zocor (simvastatin).

As a broad generality, line brand equals galenical brand development and OTC switching in the pharma world. This tactic sometimes allows pricing flexibility but more often improves the competitive dynamics a number of years after the original launch. These new dosage forms tend to allow administration to different patient types. An oral solution can greatly ease the difficulty of administration of large oral dosage forms to the elderly or pediatric populations. Another example constitutes the intravenous forms (IV), which can provide rapid loading of the product in the patients' blood stream in the intensive care setting. Even tablet development can have an impact. Melt tablets can provide an acceptable taste mask and ease swallowing of large tablets as well as increasing the chances of compliance with a particular regimen. Reducing the frequency of administration can be highly successful also. Allowing the patient to take the product only once a day versus perhaps twice or three times previously, which can be extremely difficult to manage when a child attends school.

Range Brands

The classic consumer example for range brands are the food brands such as Bird's Eye and Lean Cuisine and also the cosmetic range brands such as Clarins. Sometimes, hundreds of products that are related can be covered by the same brand name, and this allows external communication focus, clear positioning for new additions, and easier structuring of the offering for the distributor. Range brands are not as closely tied to the origins of the original (line) brand but represent another distinct extension strategy.

The pharmaceutical OTC world does have range brands but they are much more limited in their breadth and depth. The Tylenol and Panadol ranges have extended from merely different formulations of the same active ingredient (line) and now have a number of combinations that add something to the range brand such as increased efficacy due for instance to the addition of codeine in the formulation. To some extent, this strategy has worked counter to the training of one of the key influencers in the process, pharmacists. They fear the increasing chances of a dispensing mistake as a major argument to resist this type of brand tactic. Panadol is associated as a paracetamol brand, but adding aspirin components and changing the brand name to a similar sounding brand could be potentially difficult. Many patients, who were for instance aware that they are aspirin allergic, would not spontaneously check the constituents for such a well-known paracetamol-based brand.

It could be argued that the range of vaccines marketed by GSK, from their Belgium Rixensart Biologicals R&D and manufacturing site, have many of the naming similarities to consumer range brands. Nearly all of the vaccines carry the "rix" name ending (e.g., Engerix, Twinrix, Infanrix, Rotarix, Cervarix, etc., whether they are active against hepatitis A, B, polio, or the rota virus). Presumably the range benefits from recognition of the source of origin and suggests compatibility across the range. Compatibility implies they can be given to the same patient at the same time, which is a potential key determinant as many countries run national tendering systems for purchase of vaccines and organize mass vaccination campaigns.

Another interpretation of the range brand would be that it could equate unknowingly to the way "franchise" brands are understood in pharmaceuticals. The consumer theory is that a range brand focuses promotion on the range brand or on certain representative products that most effectively communicate with the consumer.

"The (range) brand can easily distribute new products that are consistent with its mission and fall within the same category."[5] Following this line of thought, franchise branding, as it is known in the pharmaceutical industry, is a poorly developed version of the range brand in consumer theory. This concept will be discussed more in depth in the chapter covering pharmaceutical franchise branding but the major differences are shown in Figure 3.4.

• Range brands FMCG	• Range brands pharmaceuticals
– An existing brand is extended to a category of products that is different from the existing category; e.g., Clarins range or the Lean Cuisine food range.	– OTC Tylenol and Panadol ranges – GSK "rix" vaccines – Franchise brands

FIGURE 3.4. Range brand extensions.

Kapferer's categorization is useful; the pharma world has parallels that are easily identifiable. The only other Rx (prescription-only brand) line or range brand variant that comes to mind are the combinations of one brand with another active substance. There are now numerous brands that over time are combined with "gold standard" generic chemical entities to create closely related brands, e.g., Capoten (captopril) was later combined with a thiazide diuretic to create Capozide in the 1990s, the distinction being that they were both used in the treatment of high blood pressure (many more examples are given in the section sustaining a brand over time). Technically speaking, they aren't a line brand because they have different chemical entities combined, but they aren't a range either, as in general they don't come in a range but are just a single product brand often with only one form.

Umbrella Brands

The first significant examples of consumer umbrella brands came from companies such as Canon and Yamaha. They are characterized by the focus of the organization being on one brand name and the use of generic names to then describe each offering, e.g., Canon cameras, copiers, printers, faxes, etc., and Yamaha bikes, pianos, guitars, and so forth. The economies of scale associated with this type of branding strategy are argued to be significant with all communication spend being focused on one single brand name across categories and around the world. This type of communication is particularly useful in categories where there is low competitive intensity; the use of the umbrella brand alone may be enough to provide differentiation in categories with low spend. The umbrella brand does not impose major constraints on the different divisions in the corporate structure.

Toshiba hi-fi can go on targeting its younger consumers, whereas Toshiba PCs can target the modern executive segment without fear of communication confusion. Experience suggests that problems arise when the umbrella brand is "stretched" too far. The image of Bic did not help its perfume brand attain success, as it wasn't credible to the consumer. Bic's problem was that it tried to stretch vertically (in terms of price and quality) when it should have been happy to stretch horizontally only (e.g., items that are disposable such as pens, pencils, razors, etc.).

The closest the pharmaceutical world comes to this is in the generic sector of the market. Following the demise of the Sandoz corporate brand with the merger of Ciba Geigy and Sandoz to form Novartis in the mid 1990s, Sandoz was reincarnated as the Novartis brand for its generics business. It was argued at the time that the Sandoz name still held considerable physician goodwill for its quality offerings, and Novartis decided to capitalize on the brand heritage to boost the recognition of its burgeoning generics division. The generic name is added to the umbrella corporate brand name. Sandoz co-amoxyclav competes against the original Augmentin from GSK.

There is an opportunity here for a corporate umbrella brand to be built in pharmaceuticals that provides the necessary reassurance to physicians and patients in therapeutic categories with low competitive pressure, where there are no promoted products in the segment. The potential downside of "stretching" the original brand into areas where it does not have credibility are probably less significant in pharmaceuticals as long as the company concerned has some relevant R&D and medical reputation.

One area that the generic houses will have to be careful about as their business model appears to be converging to that of the patented big players is whether they can stretch their brand vertically—are physicians going to find it easy to accept a new chemical entity from an organization that could only copy in the past?—Again it's the classical vertical stretch problem illustrated by Bic.

Source Brands

The one significant difference between source and umbrella is that source brands give another name to their brands, equivalent to double branding (rather than relying on a generic name). Therefore, Cacharel has different perfumes targeting different age segments (e.g., Anais-Anais, LouLou, and Eden). Nestlè is the source brand for Crunch, Kit Kat, Nescafe, etc. The source brand strategy, therefore, creates depth in the customer communication; we associate certain things with the source and others from the product-brand-specific communication. The downside in the consumer world for the source brand is that it may constrain the communication of the product brand, according to Kapferer:

> The limits of the source brand lie in the necessity to respect the core, the spirit, and the identity of the parent brand. This confines the strict boundaries not to be infringed as far as the brand extension is concerned. Only the names that are related to the parent brand's field of activity should be associated.[6]

The major pharmaceutical examples exist around branded generics. These product brands are launched after patent expiry of the original but tend to be less aggressively priced. They represent a half-way house between the patented product brands and the straight generic product as they are often promoted by a sales force in a similar fashion to the original. Rapid price erosion, which is seen when a number of generic houses compete aggressively, is not in the interests of the branded generic company as it would devalue their source brand too. This tactic tends to work best where there is a bioavailability or delivery issue that could affect how patients react to the medicine, e.g., there are a number of diltiazem branded generics due to the instability of angina in the elderly. In certain countries, this source brand strategy has seen significant success, e.g., in Germany, where companies such as Hexal continue to make progress.

Endorsing Brands

Compared with the source and umbrella brands, the endorsing brand plays a less significant role in that it offers a guarantee of the quality (scientific and technical) of the product without being the center of attention in the purchasing decision. As a result, it allows each associated brand to express itself more fully. General Motors (GM badged) endorses Opel (Europe) and Chevrolet (United States), as well as Pontiac, Buick, and Oldsmobile in a way that does not take center stage. Thus, the endorsement could be for a product brand or line or range brand dependant on the diversity of the products in question.

The endorsing brand is used in pharmaceuticals, sometimes at a product brand level (e.g., Viagra from Pfizer), and at other times as an endorsement for a therapy area or franchise; thus, Bristol-Myers Squibb Oncology "extending and enhancing human life" is used to show commitment to an area of research and development as well as an implied guarantee for quality.

How Do Kapferer's Six Strategies Fit?

Kapferer rightly points out that in reality, over time, organizations tend to end up with a mixed model and can even end up with incompatible strategies coexisting in consumer products groups. For example, L'Oreal was considered a range brand for lipsticks, an endorsing brand for Studio Line and Plenitude, and absent from Lancôme. Similarly, 3M had five levels of branding in its portfolio of domestic and industrial adhesive products. The same mixed models exist in pharmaceuticals too. Its corporate Web site in 2004[7] shows Bristol-Myers Squibb being used as a corporate brand for prescription pharmaceuticals, an umbrella brand for Bristol-Myers Squibb Medical Imaging Inc., a source brand for "Convatec–A Bristol-Myers Squibb Company," and an endorsing brand for Mead Johnson Nutritionals (minor signage for association as a Bristol-Myers Squibb company). In addition, it is used as a range brand for BMS Virology (franchise) and a brand signaling a copromotion for Plavix (clopidogrel bisulfate) with Sanofi– Synthelabo as the two company names are used together.

Generally speaking, the six brand strategies in Kapferer's hierarchy do make sense in the pharmaceutical world. Obviously, some are more heavily used than others but some kind of compatibility can be detected. Figure 3.5 expands on the original diagram, giving the obvious pharmaceutical examples at each level of the hierarchy.

OTHER BRAND EXTENSION STRATEGIES

There is at least one other brand extension strategy used in pharmaceuticals that is worth commenting on. That is the use of a single

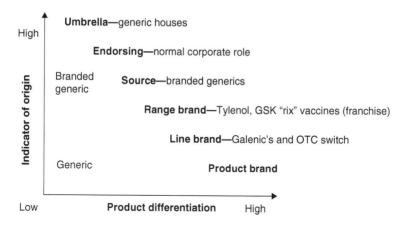

FIGURE 3.5. Pharmaceutical brand strategy hierarchy. (*Source:* Adapted from Kapferer, Jean-Noël (1991), *Strategic Brand Management,* p. 189. The Free Press, New York.)

chemical entity (and even the same dosage) with either two product brand names or alternatively in multiple indications. A limited number of examples exist, where a single prescription-only molecular entity (product) is allowed to be marketed under two names in different unrelated indications. Bupropion hydrochloride is marketed by GSK as Wellbutrin for depression and as Zyban for smoking cessation. Although this is an extension of a molecular entity, it changes the brand name deliberately. In this case, it does not correspond to a brand extension since two different brand names exist. This is comparable to the P&G experience of marketing two brands Dash and Ariel based on the same chemicals under two different positioning (whiteness and stain removal, respectively) under two different names—see Table 3.1.

Another brand extension strategy is being pioneered by the biggest companies in the sector; and constitutes the research, development, and launch of a product brand in a number of different indications simultaneously. Pregabalin from Pfizer, an antiepileptic product was launched in the European Union (EU) with epilepsy and neuropathic pain indications at the same time. In addition, it has been quickly registered in a third simultaneous indication, with the addition of general anxiety disorder (GAD). This strategy of trying to achieve launch of multiple indications at the same time is new and a direct impact of the

TABLE 3.1. Other brand extension strategies.

The same chemical entity promoted in different indications in the same geographical market

- With a shared brand name
  - e.g., Neurontin (gabapentin) promoted in epilepsy and neuropathic pain
- With different brand names
  - e.g., bupropion hcl marketed in depression as Wellbutrin and Zyban for smoking cessation

The FMCG equivalent is P&G marketing of Dash and Ariel, which are the same products with different positionings (whiteness and stain removal)

---

need to have bigger and bigger brands to replace sales of products reaching patent expiry over the coming decade.

Obviously, the resources required to be able to do this are huge and are only really available to a handful of companies in the top twenty, whose R&D spending runs into billions of dollars each year. In this case, the strategy is close to the definition of brand extension, as seen in branding theory. The regulatory authorities have realized that there are major advantages to be accrued to having two separate brand names for the same chemical entity in different indications, as it provides a great deal more flexibility for branding tactics. This tactic has been in fact quite common place in some markets (e.g., in Spain and Portugal), where multiple brands of many big sellers exist and overall the chemical entity appears to gain from the added exposure. However, what is new is the use of this tactic for potential global brands, which will be aggressively promoted in numerous indications at the same time. As a result, a number of recent requests for multiple brand names in the United States have been rejected. While the industry is under pressure, we are likely to see more deployment of this particular multiple indications strategy.

For the pharmaceutical industry, the product brand remains king, the line brand equates to galenical development and OTC switching, the range brand largely equates to OTC ranges, and the occasional Rx range in areas like vaccination and the underdeveloped brand area of the franchise. Therefore, at this stage, it is relatively straightforward understanding where pharmaceutical branding fits in compared with the first three models developed for the consumer world of branding. But, we are left with the difficulty of deciding between the technicalities of umbrella, source, and endorsing brand strategies, and how they may or may not interact with the corporate brand (i.e., corporate umbrella, corporate source, corporate umbrella, etc.), as illustrated by the BMS Web site. The source brand is best illustrated by branded generics and as the endorsing strategy plays the least role in the purchasing decision in the consumer world, it probably gets closest to the real situation in patented pharmaceuticals. Generic houses can effectively use an umbrella corporate brand. One other brand extension strategy that will be used increasingly in the future is the promotion of multiple indications and the hoped-for use of multiple brand names for the same chemical entity.

Looking at a practical level, any new pharmaceutical rep will have been through the experience of a physician or pharmacist first asking them from which company they come from and appearing to use this information as a way of deciding whether they will listen for a short time or for a long time to your sales pitch (or it may even determine if they will listen at all!). This is just a simple illustration to show there is without doubt already a complicated brand interaction that exists between the industry and its customers.

It's just that at this stage there is little brand recognition, little recognition that brands are the true capital of the business, except of course in the acceptance of the blockbuster model to provide organizational focus. Therefore, the patented pharmaceutical industry corporate brand seems to hover uncomfortably close to the endorsing brand without any real recognition of this fact—the industry just focuses on creating product brands and steadfastly refuses to discuss what endorsing advantages may accrue from a well-thought-out brand architecture, managed portfolio, and strong corporate brand (the corporate brand is looked at in Chapter 4).

Aaker Brand Drivers

The definition of different brand roles is probably what Aaker is best known for globally. He outlined five significant "Driver" roles that a brand can take (not mutually exclusive) and coined them strategic brands (including master brands), branded energizers, silver bullets, flanker brands, and cash cows.[8] A master brand is a brand that provides the point of reference and is also the focus for the majority of the investment (e.g., Toyota or GE). This terminology and the criteria for their naming are relatively straightforward and can be applied in a large part to the pharmaceutical industry.

For instance, his strategic brands are those, whose successes are vital to the organization, and this may be due to the size of the current sales they create (blockbuster equivalent), the importance of the potential for future sales (blockbuster potential possible), or they may act as a leverage point for the future (termed linchpin brands), e.g., the marketing of a niche product in a therapy area that will have the company's best future hope for big sales launched in due time.

Taking the consumer understanding of branded energizers, they are perhaps a linked product; promotion or sponsorship through its association enhances and energizes the target brand; it may be owned by the company in question or may be owned by somebody else but can be managed through commercial agreements, e.g., the sponsorship of a number of the erectile dysfunction products with national hockey or football associations in the United States. (Aaker also uses the phrase branded differentiator that improves the offering from a particular brand, e.g., a new delivery device for an asthma inhaler.)

Silver bullets are those brands that could play the most important branded energizer roles in a corporate portfolio, e.g., the acquisition of a new technology or lead product in late stage development in the context of R&D can have quite an impact; after all, the strength of a company's pipeline is known to make up a huge portion of the share price of any pharmaceutical stock. The acquisition of Celltech by UCB S.A. not only ensured that the combined company reached the top forty in terms of global sales ranking (although the Celltech sales were relatively minor), but also brought with it a silver bullet in the form of CDP 870. This biological was in phase three for Crohn's at the time of the deal and had proof of principle for efficacy in rheumatoid arthritis. This acquisition changed the future pipeline of the company from exclusively small molecule candidates to a combined portfolio of small and large molecules. Similarly, a silver bullet could be a key copromotion agreement, which gives the organization access to a superior product brand in a class the company isn't currently represented in, for example, the Lipitor (atorvastatin) deal between Pfizer and Warner Lambert in advance of Pfizer deciding for a hostile takeover.

Flanker brands are used as a direct consequence of competitor activity on an important brand, usually in a price cutting sense, where they try to nullify the impact of the new cheaper entrant or at least make sure it does not have a clear route to easy market share gains. The flanker brand has been used on a number of occasions by patented pharmaceutical companies when facing the launch of generic competition for the first time. Often via a collaboration, or less frequently as an own brand, a generic competitor is launched in advance of the first authorized generic to reach the market, so that the originator company can establish its second cheaper brand in the market well in advance and make the market more hostile to the generic competitor. SmithKline Beecham tried this tactic in a number of European countries as the patent of Augmentin (amoxycillin and clavulanic acid combination) expired. This is despite the practice having been around for years in the rest of the world. In the meantime, this legitimate patent owner tactic will continue with Johnson & Johnson (J&J), anticipated to be launching more than one authorized generic of its Concerta (once daily methylphenidate) for attention deficit hyperactivity disorder.[9]

The cash cow terminology is widely accepted following the success of the Boston Consulting Group (BCG) model. These brands provide the resources to invest in strategic, energizers, silver bullets, and flanker brands. These cash cow brands may enjoy stable or declining sales but often in the pharmaceutical context have a loyal customer base that use the brand through habit. While it's probably unrealistic to increase those cash cow sales at reasonable investment levels, they provide an important role in the corporate portfolio. Pharmaceutical companies have innumerable cash cows in their portfolios, particularly where there are products used on a chronic basis in the elderly population. This can be particularly powerful, where the nature of the patient is frail, and polypharmacy makes changes in medication unpredictable and sometimes downright dangerous. Figure 3.6 gives some pharmaceutical examples of brand roles.

Aaker Brand Architecture

Aaker uses the overall term the "brand relationship spectrum" in a later book[10] to discuss hierarchy of brands and suggests that there are four strategy options in the spectrum when deciding how to approach a new offering. A company has to decide whether it uses:

- a house of brands strategy (P&G);
- a "branded house" approach (GE, Virgin);
- an endorsed brand (3M, Ralph Lauren, Calvin Klein); or
- work with a subbrand under a master brand (Toyota Camry).

The house of brands and branded house represent opposites in the spectrum, with the endorsed brand and subbrand sitting in the middle.

Aaker contests that the house of brands strategy allows more aggressive positioning of each product brand, as no allowances need to be made for the use of the second brand (e.g., the ESPN television channel brand is not associated with its parent company Disney). This is also the chosen strategy of P&G, which allows them to differentiate on functional benefits and dominate a particular niche with a number of separate products. In the shampoo category where they hold significant share relatively anonymously with Pantene, Head & Shoulders, Pert Plus, Herbal Essences, and VS Sassoon. The product brands can even be in conflicting positions, and, in that way, P&G benefits from whichever direction the market may go in. On reflection, the house of brands strategy holds many close links with how pharmaceutical companies have behaved in the past. Often, there is very little association between the marketed pharmaceutical product brand and the parent product license holder, except where forced to do so by advertising or labeling. One pharmaceutical example of this house of brands approach is the antiepileptic drug area, which until recently was a relatively sleepy back water with low competitive intensity. Pfizer have for instance the first generation Dilantin (phenytoin), the second generation Neurontin (gabapentin), and the claimed third

• Aaker term	• Pharmaceutical example
– Strategic	– Norvasc or Lyrica
– Branded energizer	– Viagra and the NHL
– Silver bullet	– Celltech or Lipitor
– Flanker brand	– Authorized generics
– Cash cow	– The vast majority of pharmaceutical portfolios

FIGURE 3.6. Brand driver roles.

generation Lyrica (pregabalin), all as separate brands in a classic house of brands approach.

For a branded house, General Electric (GE) is the classic consumer example—the brand is GE and added to it are descriptors (e.g., GE Appliances, GE Capital, etc.). Using the house of brands side of the spectrum allows the company to decide how much individual personality they give to each separate component, whereas the branded house allows less flexibility. The advantages are that the professionalism and personality of the branded house is the focus for communication, and this has wide applicability, allowing for instance, global extension from a peculiarly British base in the form of Virgin. The equivalent in pharmaceuticals is the generic house brand. Teva, Dr. Reddy's, or Ranbaxy focus on their corporate branded house strategy when making their presentations to industry forums. The product brand is not the focus as it mostly is not given a name and relies on the generic signage as in the GE example in the consumer world.

Good examples of subbrands include the IBM ThinkPad, Toyota Camry, the TT part of the Audi TT, and the Viper part of Dodge Viper. In this way, a subbrand may allow access to a niche that would not normally be possible for the master brand, e.g., the TT adds personality and energy to the more staid Audi branding, as does the Viper subbrand when used with Dodge. The best pharmaceutical example is probably Sandoz acting as a subbrand of Novartis and so allowing access to the area of generics. Novartis is a leading creator of branded, patent-protected NCEs, whereas its Sandoz generics unit allows it access to an expanding part of the market worldwide without damaging the innovation aspect that Novartis stands for.

Endorsing brands are widely used in the consumer world (e.g., Polo by Ralph Lauren, Obsession by Calvin Klein, or 3M Post-it notes). The endorsement can be strong or token in nature, and their effectiveness appears to be strongest when the endorsing organization has credibility in the product class.[11] The area of endorsement is where most of the pharmaceutical corporate brands come into play as they mostly operate as a house of brands. Lilly "Answers that matter" is an attempt to use the corporate brand as an endorser for activities across broad therapeutic categories, for instance. This endorsing role is more deeply discussed in the corporate and franchise chapters of the book.

In reality, a large organization will have a mixture of branding strategies and the skill is about defining which brand benefits another and in what way, and then choosing appropriate actions. What is also interesting, and appears to be a contradiction, is how the pharmaceutical industry has tried to own franchises where they do business, while still operating largely on a house of brands strategy. Franchise building would appear to be an endorsing strategy moving in the opposite direction (i.e., trying to increase the awareness of the physician customer segment to the number of products the individual company has, and therefore building on expertise and experience acquired over many years—perhaps this shows how different the branding context is for pharmaceuticals). This would suggest that despite the behavior of the vast majority of the industry, deep down it knows anonymity doesn't reassure the buyer. Particularly when it is a highly technical area, where there are significant risks associated with any pharmacological choice, where advances take many years to understand, and there is the potential need for considerable physician backup if things start to go wrong.

At this stage of the analysis, it is relatively clear that many participants in the pharmaceutical industry have paid little attention to how they organize their brands, and what messages that lack of management gives to their customers. A summary is shown in Table 3.2.

PRODUCT BRAND DIFFERENTIATORS

Naturally enough, the pharmaceutical industry focuses on clinical differentiators to ensure that their product brands are competitive in the market. Those clinical differentiators translate into competitive advantage via the regulatory agency-approved labeling or via the results of significant clinical studies that increase the confidence of the prescriber due to their size and length. As a result, a huge amount of

TABLE 3.2. Corporate brand strategy examples.

Strategy	FMCG example	Pharmaceutical example
House of brands	P&G, Disney	Pfizer in epilepsy
Branded house	GE, Virgin	Generic houses
Endorsed brand	3M, Ralph Lauren, Calvin Klein	Lilly corporate
Sub-brand under a master	Audi TT, Dodge Viper	Sandoz under Novartis

intellectual effort goes into the design of carefully crafted clinical studies that are planned to show your product brand in a light at least as good as the competition and hopefully find some major differential advantage. One of the constant battles fought by the innovative patented industry is that of moving the prescriber from one heavily used product class to the next innovative class arriving via the R&D pipeline. In consumer speak, this equates to managing the category or subcategory over time.

The logic is relatively simple; your product brand may miss out on sales not because it isn't the most clearly differentiated in its category, but because the consumer chooses a different category. One easy example we have all experienced is that of choosing a car, the incredible array of different cars and their added features creates significant overlap across the categories and causes significant confusion. Why buy the BMW X3 when you can get a second hand X5 for less, and similarly sized and equipped four-wheel drives for a whole lot less? The situation is even more confusing in modern day electronics; categories that used to be diverse like computers and home entertainment systems (including televisions) look increasingly the same. In the consumer world, a branded differentiator is a brand or subbrand that creates a consumer perceived difference via a feature, ingredient, service, or program.

The same need for differentiation exists in the pharmaceutical world but tends to be thought of in a different way. For example, the choice for treatment of elevated blood pressure (hypertension) is enormous. The beleaguered primary care physician (PCP) can legitimately chose between diuretics, beta blockers, calcium antagonists, ACE inhibitors, and Angiotensin II Antagonists (nonexhaustive list). The industry has spent huge amounts of promotional money and sales force effort over the last three decades busily expanding the category, moving prescribing from one class to another, often forcing the pace, and moving on before the definitive clinical data was anywhere near available. As a result, when rational prescribing is being practiced and outcomes data analyzed, the results aren't always in favor of the most recent innovation, and class choice is a key discussion point in formulary negotiations.

This traditional product class battle ground has been centered around new chemical entities, each class being championed by an innovator; e.g., Merck (MSD in Europe) has been a major innovator over the years, leading the category development of the ACE inhibitors with the first and second once a day product brands and then launching the first Angiotensin II Antagonists well in advance of its competitors. The innovator, therefore, has to take great responsibility in positioning the class versus prior product classes, and raising the level of clinical data available to a sufficient level to reassure prescribers that safety and tolerability are acceptable (efficacy being a given in hypertension). Naturally, subclasses are established relatively quickly, one of the most significant from the patient's point of view being the frequency of administration. Capoten (captopril from BMS), the first ACE inhibitor, needed to be given twice daily and therefore was at an immediate disadvantage versus its later competitors and represented a different subcategory in the class.

There are a number of other ways by which categories and subcategories can be created in pharmaceutical classes (see Figure 3.7). These branding opportunities are more heavily practiced in the consumer world, due perhaps to the fact that many of the product brands are es-

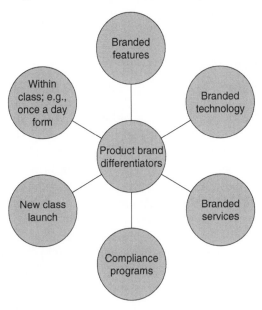

FIGURE 3.7. Pharmaceutical product brand differentiators within a category (therapy class).

sentially very similar. Therefore, there are lessons to be learned from an area where innovation is required to be other branded features, rather than the clinical route favored in the pharmaceutical world.

Branded Features

Support for an innovation is used a lot in high-tech categories (e.g., Sony i.LINK digital interface (Vaio) and BMW iDrive, which signals the onboard computer being fitted). These would be equivalent perhaps to the results from significant landmark studies, for example, the Zocor (simvastatin) 4S study that revolutionized lipid-lowering usage (Scandinavian Simvastatin Survival Study published in the Lancet in 1994).

Branded Technology

Considering popular branded technologies such as Nike Dri-Fit sports clothing, Gore-Tex North Face coats, Bose Wave Radio, and Apple iPod, the pharmaceutical equivalent would be inhaler technology, as shown in Table 3.3.

Branded Services

Examples of branded services include American Express—Round trip (corporate travel service packages), and oncology services from BMS, which even include special delivery services at one time.

Branded Programs

Branded programs include Hilton Honors, Air miles for airlines pioneered by British Airways, and compliance and retention programs for pharmaceutical medicines.

TABLE 3.3. Asthma product brands and their associated branded delivery technology.

Brand name	INN (International non-proprietary name)	Company	Branded delivery technology
Oxis	Forloterol fumarate	AstraZeneca	Turbohaler
Serotide	Salmeterol & fluticasone	GSK	Accuhaler & Evohaler
Spiriva	Tiotropium	Boehringer Ingelheim	Handihaler

Brand Alliances

Cobranding is defined as industrial alliances that are visible by the mentioning of two brand names. All alliances do not lead necessarily to the mentioning of two names.[12]

In FMCG, there are different levels of alliances, and cobranding is only one of them. The development of cobranding is a new trend in the market and has been adopted by many companies only recently. The advantages of these associations include being able to benefit from the awareness of two well-known brands, their image, their specific target market, or their technical expertise. Cobranding of ingredients has for instance become a classical nonexclusive tactic (Intel, Lycra, Nutrasweet) whereas endorsement campaigns (Ariel and Whirlpool) have been running for decades. All these cobranding agreements are linked to the need to decrease costs of development and of marketing of new products.

The concept is that two known brands will work together in developing or promoting a new product and will be visibly linked to the two existing brand names. These cobranding associations can be short term and more related to copromotional activities or long-term where both companies have long-term agreements to develop, launch and promote a new product behind both brand names. The idea is to benefit from the awareness, image, or technical skills of two equally known brands. For example, Philips and Nivea (Beiersdorf) have decided to develop and market a product brand "Philishave Cool Skin with Nivea for men." Objectives were for each of them to attract new users, enter new distribution channels, reinforce both brand images, and share development and launch costs.

As far as cobranding is concerned in the pharmaceutical world, most of the activity is confined to cobranding of corporate entities. The deals tend to focus on sharing development costs and costs of launching and promoting product brands rather than attracting new users or reinforcing both brand images. Therefore, Pasteur Aventis signals that there is Biologicals collaboration for development and supply of vaccines between the two legal entities. In this particular deal, there is some aspect of the distribution advantages as the business is quite separate from normal pharmaceuticals. In most countries around the world, distribution and administration of vaccines is via government agencies and, therefore, knowledge and expertise around how to make the systems work for you, country by country, is a vital component of success. Other cobranding of corporate entities often signals cooperation for market entry to a particular geographical market or segment (e.g., MSD collaboration for OTC medicines in Europe).

In the pharmaceutical industry, we have also seen numerous other examples of alliances, some branded as such, some not. Alliances to coresearch and develop new chemical entities are relatively common in the biotech age, with the focus for the branding activity being the financial communities who raise the funds for the biotech organizations to survive. On the commercial side, there are also numerous examples of alliances but generally the pharmaceutical industry has moved away from its understanding of comarketing towards the copromotion model. During the 1980s and 1990s, a lot of product brands were comarketed (i.e., the same chemical entity promoted under a different brand names, by different companies). It was thought that potentially doubling the resources via comarketing agreements could double the market share for the original owner of the molecule. In reality, the hard lesson was that, similar to FMCG, dilution of focus meant poorer in-market performance than hoped for. The brand development costs with the different companies and the need to establish two unique brands in the minds of the physicians led to inefficiencies and net poorer results. Despite its heritage with Innovace (Renitec in the U.S.), Merck and its comarketing partners were never as successful with the combined Zestril and Carace product brands as they were with the original.

A more common pharmaceutical tactic is copromotion—the same molecule, with the same brand name, promoted in the same territory by two companies working as separate but strategically connected partners. A good example of this is the UCB. and Pfizer relationship for the antihistamine Zyrtec in the U.S. market. UCB owns the molecule but both companies promote the brand with their own field forces sharing the revenues and profits resulting from their activities—in effect maximizing the possible promotional share of voice for the brand in the market.

Copromotion is not only used as a brand tactic but also has been pioneered by Pfizer as a strategic driver for its acquisitions. Over the years, Pfizer has entered into a number of copromotion deals with third parties (e.g., Lipitor with Warner Lambert and Celebrex with Pharmacia). The relationships acted as a form of due diligence before hostile or agreed takeover moves.

WHY MANAGE A PHARMACEUTICAL BRAND PORTFOLIO?

The traditional way to manage a pharmaceutical portfolio in the 1970s and 1980s was to focus on the R&D pipeline and discard from promotion those products that lost their exclusivity. In the 1990s, the environment for the industry got tougher as fewer NCEs were coming to market, and, as a result, more focus was given to those less numerous product brands to create the blockbuster model (which is now the accepted route to success in the industry whether mass market or specialist). Business development also really took off as an area of core competence during the 1990s, again due to the scarcity of internal late-stage pipeline candidates and the emergence of the biotech sector as a potential source of new products. Companies such as Pfizer build impressive future growth into their portfolios initially through product brand alliances (Lipitor with Warner Lambert or Celebrex with Pharmacia) and then later through aggressive takeovers of their intended target company. Despite this, and perhaps because of this, the vast majority of pharmaceutical companies continued to ignore the rest of their portfolios. Presumably, this was using the logic that near double digit growth was possible anyway, and no investment was warranted in old products due to the costs associated with sales force promotion and the nature of generic competition. Or perhaps more tellingly because the company had no intellectual energy or resources to do anything else, as marketing departments are a means to feed large sales forces, not delivering sophisticated programs to leverage brand equity. So, in summary and perhaps as an over simplification, portfolio management in pharmaceuticals tends to mean managing the big brands (hopefully capable of blockbuster status) and ignoring the rest.

Consumer companies, such as Unilever and P&G, on the other hand, have been leading the way in rationalizing their vast brand portfolios and creating growth, in tough market conditions, by providing each brand with a specific role in the bigger portfolio picture. Unilever started a process in 2000 that led to a change in focus for its entire organization; termed "Path to Growth," the goal was to focus resources on only approximately 400 of its 1,600 brands, reducing unnecessary promotional support for noncore or overlapping product brands. Those brands chosen on a global, regional, or local level would fit in with the strategic direction of the organization, would have enough media spending to be able to compete effectively, and would need to have a strong and clear positioning backed up by consumer research. The result of this review was the adoption of forty global brands (which were essentially identical globally), and 160 regional/local brands. The rest were split into various categories which included those brands that would be discontinued over time, those essentially used as cash cows, those that would be supported but whose performance would be tightly monitored, and those that should be divested. The goals of this work were to boost sales growth and improve margins and represented a bold portfolio move from one of the biggest players in the consumer world.

In theory, there should be opportunities for pharmaceutical companies to follow a similar process, perhaps with different outcomes at the end, but achieving the same objective of growth in a toughening market. Similarly, GSK has 1,400 product brands in its portfolio, which must provide enormous opportunity for strategic thinking.

There are many reasons why managing a pharmaceutical brand portfolio can pay off. From a brand perspective, there are four main areas of benefit:

- Creating synergies of activity
- Leveraging current brand assets
- Preparing the future for new products
- Providing focus for business development and alliances

Going through a portfolio review process will allow considered decision making on the desired future direction for the company and will also raise some hard questions about the business model that should be followed (e.g., long-term sales growth or long-term profit growth).

You have a portfolio problem when . . .

- The impending loss of patent on a blockbuster brand paralyzes the organization.
- Business development departments are unsure where their priorities lie.
- Product management groups are in constant battle with each other for resources.
- Differentiation appears to be impossible to achieve due to lack of investment in clinical.
- Sales targets for new product brands were never achievable, and, despite this widely known simple fact, inexpensive (no brainer) activities for less strategic brands aren't given resources.
- The organization can only focus on one promotional driver— the sales force.
- Key brands are tired, but there's no clinical or marketing energy to change this.
- Mature large brands are overresourced as emotionally the organization doesn't want to move on (and change its focus).
- Nobody cares about or manages the corporate brand.
- Life cycle management of existing brands is deemed to be uninteresting by R&D departments.
- The stock market suggests that mergers and acquisitions are the only way out.
- Finance refuses to allow cash cows to be used as trading items for globally important acquisitions or local tactical purposes.
- Individuals in senior management find it difficult to cope with understanding the therapy area complexity and customer differences.
- The corporate S&M resources don't fit the anticipated pipeline, and there's no plan.
- Sales are globally very uneven (taking into account differing market sizes and growths).
- New technologies are taking share, and the organization isn't galvanized into fighting back.

Just as a note of caution, brand structure is how the customer sees us, not the structure in which we work. Brand portfolio strategy should simplify customer understanding of the company organization and product brands, whether they are a patient, a physician, a related health care worker, or even an investment analyst.

Synergies of Activity

Big companies can be extremely complicated to work in. Just taking one example, the merger that created GlaxoSmithKline (GSK). Whereas SmithKline Beecham (SB) was a relatively focused organization from a sales point of view, relying heavily on the contributions of an SSRI (Paxil/Seroxat), an antibiotic (Augmentin), and the vaccine portfolio, the Glaxo sales portfolio by comparison was broader with less depth (other than in their asthma portfolio). Disparate therapy areas such as HIV, gastro, and antiinfectives contributed to the problem. The combined portfolio had no clearly dominant therapy class or focus and, as a result, understanding the array of different customers and their specific needs was always going to be a real challenge. Add to that a fair amount of conflicting organizational emotional baggage (where each had achieved its success in the past), and the size of the task of running a balanced portfolio from a sales and marketing perspective is immense. The problem with econometric processes is that during the data collection and analysis, many of the vital customer insights that make the real difference are lost. External consultants do draw out important considerations. If you want to maintain market share in the SSRI market, you need to have a 15 percent share of voice (SOV), and if you have to grow share then its 20 percent. But fitting it all together becomes mathematics rather than a synthesis of company knowledge gained over many years.

This type of complexity has pretty much led to portfolio paralysis, but there are starting to be some rays of hope that portfolio management is being given more attention in individual companies across the industry (e.g., the spinout from Sanofi-Aventis of twenty-two R&D projects into a new company named Novexel).[13] As part of its pipeline portfolio review, when operating as Sanofi-Synthelabo, there was a decision to divest antibacterial and antifungal fields in a structure that still allowed retention of 21 percent of the equity in the new entity. The deal went ahead despite the imminent work required to digest the hostile takeover of Aventis. There was clearly an understanding that other priorities in the bigger Sanofi-Synthelabo did not allow the focus these particular product brands deserved—otherwise they were never going to be optimized and provide their full benefit to the wider company. This is just one of a number of innovative approaches that could be taken to reduce the downsides of complexity and lack of appropriate management time and attention. Another deal announced by Sanofi-Synthelabo premerger was to spin out $1.5 billion of "tail" products and form a joint venture to increase the potential return to the organization. In the past, these new deal structures would have been confined to the biotech and R&D areas but could easily be translated to the bigger sales and marketing mother organizations as the industry starts to fully understand the negative impact of huge size and the resulting lack of agility.

Resource allocation, even in smaller organizations, remains critical to long-term success, and a clear understanding of each brand and the role it plays in the bigger portfolio is necessary to making comprehensible decisions. Often, the difficulty is deciding what the appropriate level of detail to operate at is. Everyone doesn't have to understand the intricacies of the bigger portfolio, but each person in a company should know where their efforts fit into the bigger picture. Working towards a common goal, assured that your individual contribution is wanted and needed by the bigger company is a simple way of creating synergies.

A thorough brand portfolio review can also help in two other areas. First, the creation of a clear understanding of the risks the organization has in its portfolio. Sometimes, people just get used to a particular issue always looming over them and fail to see the emerging changes in the environment that will make that feared thing come true. The eventual demise of Schering-Plough which led to the Pharmacia management team arriving *en masse* was created by a clear portfolio issue concerning Clarityn patent expiry. Second, and this one should not be underplayed either, it allows decisions on how the organization communicates both internally and externally around its brands and itself. Is Pfizer moving to being a health care provider rather than a pharmaceutical company?

Leverage Brand Assets and Launch New Product Brands

In general, the pharmaceutical industry is better at preparing for future new product brands than leveraging current assets. In part, this is due to the extended time it takes for an NCE to make it through the development process and the focus that both external commentators and internal senior management give to new product success. Extensive preparation for key opinion leaders via involvement in clinical studies and global or local advisory boards is normal, and other physician groups via medical education are now established industry practices. The maxim "prepare the product, prepare the company, and prepare the market" nicely sums up how the larger organization coordinates its activities. The lesson that doctors sell to doctors more efficiently than company people is one that hasn't been heeded enough during the days of ever-increasing sales forces, but will become more important in the future. Leveraging brands is most effectively achieved by third-party endorsement directly from one physician to another health care professional.

When it comes to leveraging current assets, the situation is less clear cut. There are many aspects to maximizing current assets, but one of the key differences, which have been discussed at length throughout this book, is the short-term nature of pharmaceutical brand promotion in comparison to the consumer world. Extending the brand life, whether that is through clever patent filings, the registration of new indications or launching well-designed new galenicals are all aspects that have to compete with the new strategic products arriving via the pipeline or alliances. In comparison to the consumer world, new product brands take years to come to market in pharmaceuticals, and, by the time a company realizes that competitors have made your brand look old and dull, it can be too late in the patent life to make the investment worthwhile. That is when product brands become poorly differentiated, nonregistration clinical groups aren't geared up to react with sufficient speed, and major opportunities are missed.

Well-conducted portfolio analysis should be able to prioritize which projects deserve investment and in which order, allowing the creation of a portfolio that manages risk over an extended period. In addition, the process should force horizon scanning for changes in customer attitudes, the growth of different markets, new technologies or the arrival of new products from competitors. In this way, a number of the brand tactics can be employed to work together effectively. For example, new product brand development allied to medical education to shape a market while employing a flanker brand to slow generic penetration. In this way, it gives the company time to react to the competitive pressures in the market.

For established brands, and in particular those approaching or already seeing generic competition, one of the most difficult jobs is to use all the incumbent marketing insight built up over many years to cost effectively maintain it. This involves acquiring a level of insight into a particular product brand, sometimes using decades of market research and usage data, which effectively crystallizes exactly why certain physicians and patients continue to use particular brands long after new alternatives become available. If those particular criteria can be understood, it gives the opportunity of very cost effective promotion, through perhaps only using one communication channel such as a patient support program. The industry struggles to believe that anything else, other than masses of reps, has an impact. Those of us who have been working with brands all our lives know that simple tactics can be very effective, e.g., in the treatment of epilepsy, dose titration is very important to be able to achieve seizure freedom. When major quality-of-life issues are at stake, such as the ability to drive a car after being seizure free for six to twelve months, then neither the patient nor physician are keen to change dosage or medication. As a result, creating the right variety of dosage forms to retain the maximum number of patients is a key determinant of brand life.

One of the real skills that is little apparent in the pharmaceutical world is knowing when to kill brands and when not to via inadvertent brand destruction. Killing brands outright in pharmaceuticals is not very common except in those circumstances where major generic competition over an extended period has reduced turnover to levels below manufacturing breakeven, and there are medically acceptable alternative treatments available for small niche indications that may be covered by the brand. Elimination from the portfolio, via sale, is unfortunately also relatively rare. A situation that is more common is when an old generic product brand has acquired so many presentations that the manufacturing complexity involved significantly hits profitability.

Inadvertent pharmaceutical brand destruction is relatively common and occurs around the time of the first major patent expiries around the world. Essentially, the company loses confidence and commitment and abandons the product brand even when there are at times compelling reasons to justify continued investment in certain regions or markets—an area that is discussed in much greater depth in Chapter 8—about sustaining a brand over time.

PROVIDING FOCUS FOR BUSINESS DEVELOPMENT AND ALLIANCES

Over the last two decades, the size of general practice (PCP) sales forces necessary to compete against the top players has increased significantly. In addition, when it comes to business development deals, for smaller commercial companies with an asset, the size and reach offered by potential business partners' sales forces is a key selection criteria. This is just one of the factors that has propelled Pfizer to the pinnacle of the industry and forced mid-sized companies to refocus their ambition away from competing in general practice. They have altered their portfolio direction in focusing their resources on specialist physicians and the product brands that they use. In specialist areas, small, well-organized players can compete effectively against the largest companies and not be at a disadvantage.

Such a company needs to have a robust enough portfolio process to ensure that it can move forward with confidence—even when some of the conclusions are unpalatable. When assessing the future pipeline potential the structure, headcount, and skill set of the future company needs to be planned well in advance. By so doing, a clear idea of the kind of external business partnerships and relationships the company is prepared to enter into is created. This also allows more productive business development and alliance activity, as the logic is clear and the speed of final decision making will be improved. The impact of these decisions will be seen in many aspects of the structure, from building particular research capabilities to acquiring a sales and marketing group with the specific specialist knowledge that will make the difference with the customer.

What Suits Your Portfolio and Business Model?

Pharmaceuticals is a high-margin business. Looking at high-margin businesses in general, they tend to compete on quality rather than quantity. As a result, they have to invest to maintain or increase that quality, or else the margin will erode over time. To do this, they tend to have a high cost base and at times have to be prepared to lose volume. This type of business model suggests that the measure of success should be return on capital employed (or return on equity)—one of the measures used by stock market analysts. So far, pharmaceuticals have driven the success of their business model through increasing the top line sales year on year. The necessity to do this has driven up the costs of sales and marketing to unprecedented levels, levels which are universally accepted to be too high for the long term as they have already reached saturation point in many major markets, e.g., the number of medical reps in the key U.S. market. The problem for the industry is that there is likely to be a first mover disadvantage, i.e., the first company to rationalize their sales and marketing spend will lose out competitively to those rivals who maintain their expense and are desperate to show it was the right choice. The decisive change factor may therefore turn out to be an external factor, such as the imposition of price controls in the U.S. market (at present unlikely under a second Bush administration).

If the measure of success in the pharma sector was moved to return on capital employed, then different behaviors would be rewarded over time. Portfolio brand management would be one of the essential tools to achieving success in the new world.

CONCLUSIONS

The strategic discussion around brand hierarchy came from simple beginnings, when brands signaled the origins of a product. This subject is probably the most relevant for pharmaceuticals due to the vast merged portfolios which lack scope, roles, and interrelationships that work together.

The pharmaceutical industry focuses on product branding over either franchise or corporate brands. In the future, it needs to make the intellectual leap of faith to restoring the industry brand and establishing strong corporate brands, so that franchise and product brands can deliver value.

Kapferer identified six models for brand strategy and maintains that significant thought has to be given to the overall structure of each portfolio. Over time, organizations end up with a mixed model of the various brand strategies. Each branding strategy has an equivalent in pharmaceutical branding:—product brands; line brands (galenical forms and OTC switches); range brands (OTC ranges and the franchise brand); umbrella (generic houses); source brands (branded generics); and endorsing, which was found to be the most common corporate activity in the pharmaceutical sector. In addition, one other strategy was identified, where multiple indications for the same brand or multiple indications for different brand names are utilized.

In his explanation of branding strategy, Aaker names five significant "driver" roles that a brand can take. Each one similarly has utility in pharmaceuticals, e.g., strategic brands (blockbuster or potential blockbusters), branded energizers (E.D. sponsorship deals with U.S. sporting federations), silver bullets (Lipitor copromotion for Pfizer in advance of the hostile takeover of Warner-Lambert), Flanker brands (in-house branded generics launched in advance of patent expiry), and cash cows (the vast majority of any top pharmaceutical company's portfolio).

As far as hierarchy is concerned, Aaker identified four strategic options—all of which are used in the pharmaceutical world. At this stage, the "house of brands" strategy predominates, whereas the "branded house" strategy is used by the generics houses, the "subbrand" by Sandoz acting as a subbrand of the innovative Novartis master brand, and, finally, the endorsing brand, which is often used (but rarely understood) in pharmaceuticals.

Portfolio brand management is at a rudimentary stage in the pharmaceutical industry—there are numerous signs and symptoms that show when a company has a portfolio problem. The advantages of portfolio management include, among others, benefiting from synergies of activity, leveraging brand assets and preparing for new launches, providing focus for business development, and deciding what suits your business model. If the measure of success in the pharma sector was moved to return on capital employed, then different behaviors would be rewarded over time. Portfolio brand management would be one of the essential tools to achieving success in the new world.

Chapter 4

Corporate Brand

Within the context of this book, the corporate brand is just one level of the branding hierarchy. Use of the corporate brand is common within both classical consumer brand theory and the somewhat different perspective of branding in pharmaceuticals. The consumer argument is that corporate branding does make a difference, and to some extent every shopper knows that. The modern shopping world is increasingly made up of "grey" goods or fake goods whose origin is unknown; the establishment of trust as well as differentiation for the product brand is becoming more important within the consumer world. Market research shows that image, purchase intent, and sales revenue are closely related, but it is difficult to pin down exactly what part of that image the corporate brand can be credited with, as the interactions are numerous and separate. This leaves us with one thing that is easy to believe, and to quote J. Gregory from the *Journal of Brand Management:*

> Image has been shown to affect not only sales volume but also the price that customers will pay for the products and services offered.[1]

This might be believable, but it does not provide many of the answers. Some progress has been made in the last five years concerning quantifying the benefit of corporate brand communication, most notably in the research conducted with financial influencers and corporate decision makers. Improvements in corporate familiarity and favorability were clearly linked to improvements in stock price. This was attributed to first, when all other things were equal, an improvement in image leads to sales growth, and second, that the same image

Pharmaceuticals—Where's the Brand Logic?
Published by The Haworth Press, Inc., 2007. All rights reserved.
doi:10.1300/5836_04

improvement leads the stock market to give a higher price earnings and cash flow rating. Not surprisingly, the top five companies in the US 2001 research were Coca-Cola, Microsoft, General Electric, Disney, and Johnson & Johnson (first to fifth), each a consumer goods company with clear corporate brand strategy and each of which invests significant resources in corporate brand building.

Another measure of corporate brand strength is provided by an annual "Interbred" and "Business Week" collaboration, which over the years has tracked brand valuations. This analysis places three big pharma brands in the top 100 in 2004.[2] Those three were Pfizer, Merck (pre Vioxx withdrawal), and Johnson & Johnson (J&J), but on this occasion J&J had dropped in valuation to a lowly eighty seventh place. The valuation method again mimics how financial analysts value companies, but within the calculation intangibles such as patents and customer convenience are stripped out of the figures. This may explain the plummeting rating for J&J and presumably also why Pfizer only had a 2004 brand value of $10.6 billion in 2004. This appears strange when just one of their product brands (Lipitor—atorvastatin) sold $10.9 billion on its own in 2004 according to Pfizer accounts, and you would imagine that kind of financial sales performance would be represented to a large degree in the valuation of the corporate brand.[3]

From a financial market perspective, some investment in corporate brand building for pharmaceuticals is a worthwhile investment. This perhaps explains why at this moment there is some limited activity in this area. But there should be more depth to what a strong corporate brand could bring.

In digging a bit deeper and trying to find pharmaceutical-specific truths, things that intuitively seem right for those who have worked in the industry, there is some limited market research on the subject with physicians and pharmacists. The early research was reviewed by Wright and Fill in 2001[4] with a number of general assertions being possible: as expected, there is a link between increased prescribing and positive company image and that, when given a choice between very similar brands, physicians can base their selection on their opinion of the company involved. According to the review of three major pieces of research with physicians in 1993 (United States), 1997 (Greece), and 1999 (United Kingdom), the attributes that made up that image were (not necessarily in order):

- drug effectiveness;
- level of R&D;
- knowledge demonstrated by the sales representative;
- sensitivity to pricing concerns;
- educational orientation; and
- efforts to train staff.

These attributes appear very "self" focused when looked at within the bigger picture and perhaps would be different if the market research had been conducted with specialist physicians who are involved increasingly in global key opinion leader networks or other health care professionals. What the attributes don't explicitly say is "trust" (perhaps the above imply this however). Everyone who works in the industry knows that effective and safe product brands are vital and that the person facing the customer (the rep) has a huge impact on what the customers think about the company. Health care is a trust business in which there are huge unknowns and major risks are taken by patients and physicians. Perhaps we are missing this implicit point and, as a result, underplay the significance of positive benefits of the R&D conducted, failing to make enough effort communicating that significance to the physicians and their related health care workers via the corporate brand. Having said that, the list does make sense from a physician's view point but what is also apparent is that the research or knowledge of this area isn't keeping pace with the changing environment. The advent of DTCa, cost shifting to patients (in all markets), major OTC switches, the Internet, and major structural changes in how health care is delivered across the world makes these results appear minor and noninclusive.

The world is becoming increasingly aware of the risks of fake medicine production and distribution. What appears to be happening is that traditional illicit drug dealers are figuring out that their trade is both targeted by law enforcement agencies around the world and dangerous. To deal drugs, you have to do business with drug addicts at some point in the distribution chain, and this is inherently hazardous. If, however, you get into the fake medicines business, not only is active pharmaceutical product (API) freely available over the Internet, but after you have manufactured and packaged the fakes, the people you are going to deal with aren't drug-crazed maniacs, but duped pharmaceutical distribution people. So far, this industry doesn't require anyone to carry a gun!

When assessing illegal activities, the financial returns across different types of business models appear to give very different levels of return on investment (ROI).[5] Heroin smuggling can give a very reasonable ROI of approx $20,000 for every $1,000 invested, while the safe business of counterfeit credit card fraud tends to give a return between $2,570 and $7,000. Counterfeit currency merely returns $3,000 (based on Asian wholesale value giving 30 percent of face value), but when it comes to counterfeit medicines, the business model ROI changes dramatically. For every $1,000 invested in fake Cialis at the active ingredient stage, the estimated uplift to retail return can be more than $500,000! Which business would you prefer to be in? Personally, I would choose the one with the highest return for the lowest personal risk (see Figure 4.1).

Increasingly, physicians and particularly patients are asking questions. Who are these guys who make my medicine? Are they legitimate? How do they ensure my pills are the real ones? What is their heritage? Do they know what they are doing? What is their culture? What are their values? Will their supply systems make sure I've always got my medicine? And do they have loyalty programs for me? A lot of this adds up to—can I trust them?

The issues involved in the twenty-first century are much bigger and intuitively we know that the corporate brand is now having an increasing impact on patients, health care workers, and also our own

Business model

Activity	Return
Heroin smuggling	$20,000+
Credit card fraud	$2,500-$7,000
Counterfeit currency	$3,000+
Counterfeit medicines	$500,000+

- Drug dealing is highly dangerous
- The least hazardous business to be involved in is pharmaceuticals distribuition
- No criminal contacts required
- The trade is not heavily target by law enforcement agencies

FIGURE 4.1. $1,000 invested in illegal activities. (*Source:* Adapted from oral presentation of Steve Hawgood.)

lives at work and in the communities we live in—but as usual we don't really have a pharmaceutical brand model to explain it. The stakeholders for communication are much wider than we have traditionally thought of; it is no longer merely physicians, pharmacists, and related health care workers, but patients, their relatives, the financial community, health care administrators, government civil servants, elected representatives, etc.

This section of the chapter will therefore look at the corporate brand in more detail without being exhaustive—largely because there don't appear to be definitive answers for pharmaceutical companies, so we have to be led by gut feel and consumer branding knowledge. The architecture section has already looked at the various ways a corporate brand could be used and should be referred to for strategic purposes.

WHAT DOES A CORPORATE BRAND STAND FOR?

The corporate brand represents many different things to different people and is more complicated in that aspect than merely managing a product brand. Done well, the corporate brand can be leveraged across a wide range of activities, pulling together all the different products brands and services that an organization may offer, creating differentiation from similar products or services and creating trust for the purchaser. Some of the major things that a corporate brand stands for are set out in Table 4.1.

Heritage

The heritage of an organization has been underplayed in pharmaceuticals due to the house of brands strategy chosen by virtually

TABLE 4.1. What a corporate brand stands for.

Heritage
Values, strategy, and priorities
Who works there, what is their expertise, and how they will help
Global citizenship
Investment strength, size, and respect

everyone. Few pharmaceutical product brands have built a long-term relationship with customers like, for instance, Aspirin from Bayer; the vast majority stands alone and don't even form part of a franchise or range. They, therefore, can't communicate the many decades of research that a company has had to go through to discover a new chemical entity and bring it to market via formulation, manufacture, clinical development, and registration. A product brand can't emphasize the research technologies that have had to be mastered, the development difficulties encountered, or the level of knowledge required of the patient and physician environment to create an innovative family of compounds—one of which eventually made it to market. Heritage, therefore, represents a history of expertise that is often missed in communication with the outside world and is also underplayed by consultants who think that the highest price is always going to get the best scientists. This isn't the case. Many scientists are more like advertising agency creatives, They want certain things in life (like intellectual freedom) more than the big bucks the management consultants think will always be enough to secure their services. Pharmaceuticals is an intellectual business; although we sell pills, we are really selling the intellectual scientific knowledge that goes with it, i.e., safety, efficacy and side effects, its interactions, and how it can be combined with other pills safely.

Values and Priorities

The corporate brand also represents the values and priorities of the business (i.e., what is acceptable behavior within the culture). How the values of the corporate brand then interact with strategy to give priorities is vital to the good health of any organization, but particularly in a long-term business like pharmaceuticals. Is the company prepared to invest for the long term, or is it short term in its outlook? Will it regularly change its mind about investment priorities as it is driven by the stock market? Where does it stand as far as business ethics is concerned? Is it prepared to invest in landmark studies, or does it just follow the innovators? Does it make cynical business decisions or take into account how customers will perceive its behavior over time—does it care? A cynic would easily remark that the pharmaceutical industry is not interested in peoples' health, but merely in the profits that can be made. The CEO and the Board have to have the necessary skill sets to allow management of the brand at their level rather than merely delegating it to someone in a staff function while driving the sales and profit delivery.

Corporate Personnel

The cultural aspect brings us to the next expression of the corporate brand—the people employed who represent it. Due to the technical nature of the work, having highly qualified people involved in drug discovery, development, and marketing is vital to success. The research discussed earlier concentrated on representative relationships, but in reality much of this aspect is led from the center of the organization. To build trust around a product brand offering, it is essential to have the necessary expertise available within a company to reassure the key physicians that they can use the product within a safe environment, e.g., does the company take adverse event reporting seriously? Does it have the required level of expertise in house to interpret that data accurately? And will it take action quickly enough if a problem occurs?

Global Citizenship

The corporate brand also stands for global citizenship. Protection of the environment is becoming a much more sensitive issue, and this is only likely to increase in time with the refusal of the United States to ratify the Kyoto treaty and China exerting its industrial might. Although the environmental impact associated with manufacture of pharmaceuticals is minor in comparison to say the oil and gas industries, the issues remain important. Good corporate citizenship for our industry is not only about meeting the required environmental standards but also being able to comply with the ever more stringent worldwide regulations. Delays to the launch of a new product brand, for instance, due to poor manufacturing compliance, is a very public way of showing poor corporate citizenship. How corporations behave in local markets, when increasingly they are operating in a global environment, also is part of the good citizenship. Does the company merely repatriate as much of the profit as possible without investing in the local community, or does it look after its people and where they live? Philanthropic acts are also a part of this aspect of corporate brand building; creating programs that benefit the community or giving donations at times of international crisis are good examples. My own perception of the industry's response to the Tsunami disaster on December 26, 2004, was that it was rapid and significant, and this message got across on global news networks. There was also follow-through from various big companies with a mixture of medical supplies and cash donations, making the industry effort significant.[6] Perhaps the big companies are starting to take more care in how they are perceived, and this can only be a benefit for the industry brand and individual company brands moving forward. Trying to create good feelings about a company, rather than merely accepting that everyone hates you, has to be good for both internal and external stakeholders.

Investment Strength

Last, but not least, the corporate brand represents strength, size, and respect within the investment communities. Very few pharmaceutical companies are privately owned, and, as a result, the industry has to work within the confines of the stock market, quarterly results, and related cross sector comparisons. A huge proportion of the pharmaceutical stock price is related to the perceived strength of the pipeline, and then comes the business development ability, as well as the aptitude to maximize an asset from a sales and marketing point of view. Companies that can demonstrate these abilities have a different rating to those that cannot. There are instances of hugely successful corporate brands such as General Electric that provide an aura of success across widely differing industry sectors. As yet, the pharmaceutical industry hasn't needed to create strong corporate brands that can compete within the consumer world (with the apparent exception of J&J), and the benefits are therefore untested.

RELEVANCE OF A CORPORATE BRAND IN PHARMACEUTICALS

The easiest answer, concerning whether a corporate brand in pharmaceuticals is relevant at this stage, is—unfortunately not much!

The exception may be for the investment community. The heritage aspect of the corporate brand is often only exercised to give yet another field force a name, while the other aspects of the corporate brand covered above are given some lip service but no real focus. The corporate brand needs to be managed at the top of the organization, and it is therefore a sad indictment that the most frequent use for the corporate brand comes in field force naming. Even this can appear extreme when a single entity such as GSK has over 8,000 representatives in the U.S. market alone (9,000 in Europe and 11,000 internationally). As a result of all the mergers over the last two decades, there are many names to choose from but, once you already have ten separate primary care physician sales teams, then finding yet another name becomes still difficult! Unfortunately, when the leaders of the industry appear in the press, they are usually there for the wrong reasons; whether it is to refute the excesses of their pay packages (J.P. Garnier, GSK) or to defend the withdrawal of a product brand and the circumstances around it (R.V. Gilmartin, Merck).

Brand commentators such as Tom Blackett express the opinion that, although the pharmaceutical industry is not good at corporate branding, this is not a surprise because it isn't very good at product branding either![7] To back up this premise (probably correct), he cited the annual league tables created by Interbrand of the world's most valuable brands in 2002. At the time, only two pharmaceutical companies made it into the top 100, Pfizer at twenty-eight and Merck (MSD) at thirty-three. Merck managed to maintain a comparatively high ranking for pharma companies until 2004, but the Vioxx withdrawal will no doubt affect the standing of Merck when the rankings for 2005 arrive. Up until the Vioxx withdrawal in 2004, Merck had been a financial bellwether for the industry over a number of decades. Its decision to withdraw a $2 billion plus product brand came as quite a shock to both the industry, financial, and consumer worlds. In 2004, the most valuable brand was (no surprises here) Coca-Cola—a company whose major contribution to a better world is to make sugar-flavored soft drinks. There is opportunity here if only the industry is prepared to take the issues on board and think about the value a well-developed corporate brand can deliver.

Looking at a practical level, any new pharmaceutical rep will have been through the experience of a physician or pharmacist asking which company they come from. They appear to use this information as a way of deciding whether they will listen for a short time or for a long time to your sales pitch (or it may even determine if they will listen at all!). This is just a simple illustration to show there is without doubt already a complicated brand interaction that exists between the industry and its customers. This observation is, however, complicated by the relative success of contract representatives— surely being from a completely unknown company should impact on their performance. The fact that this doesn't appear to be the case suggests that the customer has become used to the excesses of the industry where escalation of rep numbers has been common practice and further undermining of the corporate brand is inevitable.

LEVERAGING CORPORATE BRANDS

According to those that specialize in consumer branding there are numerous ways in which a strong corporate brand can help a business, and this is probably demonstrated by those companies that invest in this area in the wider consumer world. As with any brand, a well-managed corporate brand should be able to differentiate the company from its competitors' companies; Virgin and General Electric have achieved this. Table 4.2. sets out how the corporate brand can be leveraged.

Increasingly, governments and major health care providers are requiring companies to negotiate at a local, regional, or national level

TABLE 4.2. Leveraging the corporate brand.

Differentiates in negotiations and when launching new product brands
Provides reasons to purchase above the product brand; e.g. loyalty programs and therapy areas with low competitive intensity
Projects credibility and heritage
Allows easier geographic expansion
Helps external communication
Supports internal values and strategy implementation
Can be leveraged in M&A and business development
Helps attract the right talent

when it comes to pricing and service contracts, e.g., the state of Florida has gone into agreements with a number of big pharma companies in an attempt to reduce costs and cut a good deal for certain population segments. In this sense, a corporate brand can energize a discussion, add credibility to a position taken in the negotiation, and perhaps open doors for formularies that other wise would have been closed. A strong corporate brand can put a negotiator in a good position because their particular company has a reputation for delivery and service. Naturally, this cuts both ways—those companies that prevaricate and negotiate in a cynical fashion rapidly become known to the health care providers in question, and a lot of damage can be done quickly.

A strong corporate brand can also be leveraged when launching new product brands around the world, giving reasons to use above the product brand by projecting heritage and relevant therapeutic credibility. The previous track record of the sales and marketing comes under scrutiny, and this can play a significant role with not only physicians and associated health care professionals but also industry and financial commentators. The ability to maximize the opportunity of any particular product brand given the limitations of its clinical data and labeling is a real asset. Novartis has done well with Crestor despite what appears to be persistent differences of opinion about the clinical evidence within the medical community.

Individual product brand loyalty programs can be extremely difficult to control. What started as an obvious marketing idea to help retain patients on therapy via patient information, coupons, or discounts can rapidly turn into a monster that devours huge slices of the marketing budget. Everything starts well, and then you find out you have to move the data house due to the volume, then you realize that there are lots of associated costs to the program, which are burning the effective spend, and it is difficult to retrench without inevitably affecting the loyalty you set out to improve in the first place. This is a perfect opportunity to leverage a single branded program across the portfolio and bring the patient closer to the company in question. Using the corporate brand as an endorsing or umbrella brand in classic consumer tactics could effectively package a number of schemes together and provide benefit not only to the product brand but also the corporate brand. In this, both brands add value to each other.

Whether or not the world's regulatory authorities like it, DTCa is here to stay in the United States. Its leakage around the world will increase the phenomenon of patients learning more about their medications in every market, due at the very least to the increasing use of the Internet globally. Even in the United States, it is extremely difficult for a prescription product brand to accurately communicate with a consumer audience due to the complexity of regulations that persist—frankly, the industry has been making some lousy adverts that appear capable of confusing everyone. The corporate brand, therefore, represents a real opportunity to take relationship building to another level, allowing easier geographic expansion. The focus of spend will be nonpromotional, and a longer-term plan can be implemented. It is an ideal forum to build trust with patients concerning the medicines they take on a day-to-day basis, and fits in with the need to start managing the vast portfolios of product brands that each top company holds. In noncompetitive areas, in particular, a strong company brand can make the difference when choosing between very similar product brands. In addition, the issues around some of the well-publicized problems associated with counterfeit goods. The baby milk scandals in China provide an obvious leverage point for a change in strategy.

The traditional use of the corporate brand is associated with shareholder scrutiny. There are three broad aspects with external customers that include ensuring corporate social responsibility is adequate, that financial analysts provide relevant ratings which maintain the share price, as well as the credibility of the brand when it comes to mergers and acquisitions. Corporate social responsibility is becoming more relevant in the modern world, particularly after Enron and WorldCom. This extends beyond the purely financial aspects of doing business; the industry is under direct scrutiny not only from governments but also from nongovernment organizations (NGOs) to improve access to medicines and not just in poorer countries. In addition, they are creating pressure around acceptable pricing strategies that make a single global price less easy to defend, given the politics inherent in health care provision and each individual country's ability to pay. GSK has been at the forefront of corporate conscience advertising with their "Science with a conscience" campaign in the U.K., with provocative questions such as "Sarah says your company exploits the NHS, mum" and the subsequent defense of the company's position.

As far as the financial markets are concerned, research on the "beta factor" (a financial instrument of measurement used within corporate banking groups) has shown that reputation is a key component in share price stability.[8] For instance, an assessment of an individual company's share price movement versus movements in the market, as a whole, show that the average pharmaceutical company is significantly less prone to under- or overreaction than the average biotech company.

The pharmaceutical industry has seen consolidation through agreed mergers, but has also experienced a number of hostile activities in the last decade. The acquisitions of Warner-Lambert by Pfizer and Aventis by Sanofi-Synthelabo occurred relatively recently, while acquisition of major share tranches is another form of hostile activity demonstrated by Novartis' holdings in Roche. A strong corporate brand that has a previous track record of acquiring and then extracting the value from a deal is a powerful advantage when it comes to either friendly or hostile activities. Financial credibility is the key to success as it has been extremely difficult to prove the worth of many mergers over the last decade. Therefore, apart from the investment bankers who benefit from their fees, there is a certain amount of skepticism about the value of the consolidation, in shareholder groups and even among the people who work in the industry. As one skeptic remarked at the combined R&D pipelines of Glaxo Wellcome and SmithKline Beecham when forming GSK, "If you take two empty boxes and put them together, all you get is a bigger empty box." The situation is never simple, however, as demonstrated in one of the deals. A corporate brand that had managed to make headway despite the continuous merger activity of the 1990s, such as Aventis, should have been capable of repulsing the overtures of a smaller company with more limited merger experience. In this case, French politics probably played a large part in the final result and perhaps will leave top Aventis executives who lost out in the process wondering whether a stronger corporate brand strategy would have provided more protection.

Licensing and alliances are a key element within the industry model that can be enhanced by a strong corporate brand. Some companies manage to make alliances work consistently, while others struggle because their corporate culture finds it difficult to accept compromises or adaptation to local ways of operating. The balance of licensing has changed over the last decade, with the power shifting onto the biotech side of the bargaining. If a biotech now has sufficient funds, there is no need to partner early and can even go alone if the corporate brand and the key personnel have sufficient credibility. The deal structures available have also changed where biotechs can now charge for options on the product brand or open an auction to ensure that the highest price is achieved. The chances of partnering are increased by the flexibility shown by large corporate organizations in the structure of the deal and how previous partnering experiences have progressed when a product brand makes it to market. Alliances are also created between pure pharmaceutical players, when either their geographical coverage or expertise don't match the best way to optimize an asset. Johnson & Johnson MSD Europe was a nonprescription joint venture that was aimed at growing the MSD OTC range and promoting new potential switches. Eventually, it was bought out by J&J in 2004, but represented a good example of cobranding of the two corporate brands. In the field of vaccines, MSD entered into another innovative European structure in its Aventis Pasteur MSD joint venture, which became Sanofi Pasteur MSD in 2005.

A strong corporate brand also helps in making decisions concerning structure and naming at the time of mergers and acquisitions. The relative brand strength of the numerous company brands, be they corporate, franchise, or product brands, need to be assessed for their strength and what they convey to the relevant target customers. There could well be credibility for one brand in one area and a complete lack in others. When Pfizer took over Pharmacia, there was clear heritage for Pfizer in human and animal health, while it lacked credibility in diagnostics. As a result, Pharmacia diagnostics was retained after acquisition, and is not associated with the corporate brand. Novartis was created in 1996 by the merger of Ciba-Geigy and Sandoz; a number of years later, in 2003, it resurrected the Sandoz corporate brand to position its generics business clearly while benefiting from the heritage and scientific credibility of the old Sandoz corporate brand. Research had shown that the name still retained a strong reputation and high level of awareness with physicians, pharmacists, and even patients and was used to cover the acquired businesses of

Geneva Pharmaceuticals in the United States, Azupharma in Germany, and Biochemie in Austria.[9]

Talent attraction and retention is an important critical success factor for any organization, and this is particularly fierce within the R&D area, where pay is often not the primary motivator of the individuals who will deliver the results. In this sense, flexible packages and encouragement of creative and original thinking is an important part of the environment. This aspect of productivity was recognized at the GSK merger and lead to the setting up of innovative centers of excellence that allow like-minded scientists' rapid decision making and rewards linked to results. Again, a strong corporate brand that expresses the right kind of values will help in the recruitment process and the retention of talent over the long run.

One specific market that has seen numerous business models employed is Japan. To initially break into the second biggest market in the world and allow access to the world's oldest population, Western pharmaceutical companies have had to employ a range of strategic approaches, including joint ventures as well as acquisitions. The success rate, it has to be said, is variable, with some companies running virtually "shadow" Western organizations next door to their acquisitions and others happy to allow their new assets to maintain an independence in the name of the subsidiary (e.g., Banyu-Merck).

Managed well, a corporate brand can be a differentiator in its own right, can be leveraged in mergers and alliances, can energize and provide credibility, can provide trust, can support company values, and can encourage employees to think outside the box. It can facilitate the ultimate branded house strategy and the chance to maximize branding opportunities in a rapidly changing pharmaceutical environment.

DOWNSIDE OF CORPORATE BRANDS

The potential downside to association of the corporate brand with its product brands has always been the risk of contamination, one product brand failure affecting the rest of the company. One of the key criteria behind the branding strategy of a house of brands is when the corporate name is rarely used and each product brand stands on its own merits in an attempt to limit any potential damage. The negative associations of a significant product brand withdrawal are well known within the pharmaceutical industry and, if not managed with skill and great care, can represent a hangover for the entire organization over an extended period of time. Wyeth suffered litigation that has continued over many years following allegations of damaged heart valves due to its diet product brand Fen-phen. Despite setting aside $17 billion to cover Fen-phen legal costs and liabilities, extra funds may well still be required to settle all cases.[10] The long-term effects on the corporate brand have been significant, with just one example being the ring fence it sets around the company when it comes to participating in mergers and acquisitions. The risk profile associated with the outstanding litigation means that the Wyeth has been effectively barred from many transactions.

Other product brand withdrawals have had less impact on the fortunes of the corporate brand in question. The withdrawal of the anti-diabetic product brand Rezulin (troglitazone) originally marketed by Warner Lambert did not give the acquiring Pfizer a long-lasting taint in the market. In a similar fashion, neither did GSK's withdrawal of their IBS product brand Lotronex. This may well be associated with the relative size of the financial issues involved but is also significantly associated with how the withdrawal occurred and how the company was seen to react at the time of the bad news.

Corporate Brand Building

For an industry that doesn't possess a great deal of brand expertise and has consigned the corporate brand to something necessary for the financial markets, positive change will not be easy. Simple questions such as what brand image should be projected are quite complex, considering the vast array of audiences that need to be covered. Making the corporate brand and what it stands for relevant to both internal and external customers is difficult especially as the value propositions are quite different. Those that manage the corporate brand need to find a way to manage the brand in different contexts in a way that supports organizational strategy, avoiding if possible the most visible negatives and come up with clear organizational rules about how and when to leverage the brand. Being confident that synergies are possible and that corporate brand activities improve R&D and sales and marketing (S&M) credibility will take a long time to build but change needs to be signaled. After all it's a trust business.

CONCLUSIONS

There appears to be a link between corporate image, related revenue, and the price that customers are prepared to pay within the consumer world. In comparison to consumer corporate brands, pharmaceutical brand valuations are low.

On the pharmaceutical side there is a link between increased prescribing and positive company image and when given a choice between very similar brands, physicians can base their selection on their opinion of the company. Health care is a trust business, a business where there are huge unknowns and major risks are taken by patients and physicians. Perhaps we are missing this implicit point and as a result underplay the significance of positive benefits of the R&D conducted, failing to make enough effort communicating that significance to the physician and their related health care workers via the corporate brand.

Awareness of counterfeit medicines means that everyone is going to be asking questions. Who makes my medicine? Are they legitimate? How do they ensure my pills are the real ones'? Will their supply systems make sure I've always got my medicine? A lot of this adds up to—can I trust them? This provides a leverage point for a change in brand-building strategy.

The corporate brand stands for heritage, values, strategy, and company priorities. It also reflects who works there, their expertise, and how they will help. The accepted pharmaceutical role is that of signaling if the organization is a good global citizen and providing a hint as to the general state of its financial strength.

Consumer research suggests that strong corporate brands give significant advantages. They can differentiate on their own in negotiations and, when launching new product brands, providing reasons to purchase above the product brand (e.g., loyalty programs project credibility and heritage). In addition, they allow easier geographic expansion, help external communication, and support internal values and strategy implementation. Those strong brands can be leveraged in M&A and business development, while helping attract the right talent into the organization.

The potential downside to association of the corporate brand with its product brands has always been the risk of contamination, one product brand failure affecting the rest of the company. One of the key criteria behind the branding strategy of a house of brands is when the corporate name is rarely used and each product brand stands on its own merits in an attempt to limit any potential damage.

Brand commentators express the opinion that, although the pharmaceutical industry is not good at corporate branding, it isn't very good at product branding either. A cynic would easily remark that the pharmaceutical industry is not interested in people's health, but is interested merely in profits. For an industry that doesn't possess a great deal of brand expertise and has consigned the corporate brand to something necessary for the financial markets, positive change will not be easy. Simple questions such as what brand image should be projected are quite complex, considering the vastness of the audiences who need to be covered.

Chapter 5

The Franchise Brand

The basic meaning behind the word franchise is very different when looking at a consumer marketer's or pharmaceutical marketer's understanding of the word. Within the consumer world, the word "franchise" is not widely used within the context of brands. An organization may make use of a franchise business model, often with brands attached to it such as McDonald's or Starbucks, but it does not have the same depth of meaning that it holds within pharmaceuticals. The term "pharmaceutical franchise" is used to signal a group (or range) of product brands that target a similar indication or physician segment (from a single company) with the objective of leveraging the brand equity inherent from previous (or future) product brand offerings.

> A range brand platform is created by a range brand (a brand that spans product categories) whose identity includes a differentiating association that is applicable across categories. Creating a point of differentiation that can be employed across a set of target product categories is very different from creating one that will resonate with a single candidate product category.[1]

Examples of range brand platforms are Volvo (leveraging safety across its range of vehicles) or Tylenol (leveraging the efficacy of acetaminophen across cold, flu, sore throat, menstrual pain, etc.). Creating a range brand platform is a strategic move covering groups of categories and takes a long-term view; the brand scope can be enlarged over time but it is a different proposition from product brand

Pharmaceuticals—Where's the Brand Logic?
Published by The Haworth Press, Inc., 2007. All rights reserved.
doi:10.1300/5836_05

creation. It offers coherence and structure, allows each product brand to have its own identity, at the same time it leverages strong brands. It thereby reduces brand building costs while the range offers reassurance to the consumer that the firm has the necessary expertise to deliver on the franchise brand promise.

Reading the theory and equating it to pharmaceuticals suggests that the term "franchise" within the pharmaceutical world is roughly equivalent to a poorly thought through consumer range brand platform. The consumer range is an existing brand extended to a category of products that is different to the existing one (e.g., Clarins cosmetics ranges). But first let's take a look at franchise branding in more depth.

PHARMACEUTICAL FRANCHISES

The pharmaceutical global market is a fragmented one. Despite Pfizer reaching a double-digit global market share with its acquisition of Pharmacia, the world pharmaceutical market is made up of thousands of niches. This is unlike many consumer markets where consolidation has left the major players with 20 percent market share or more. Although the top pharmaceutical companies have been going through a merger and acquisition cycle for two decades, the same kind of dominant market shares are unlikely to be seen. Some of the reasons for this are the complexity of each disease, major interpatient variations, and the different ways patients need to be treated as a result of this. Largely speaking, neurologists aren't bothered by who has the highest global market share with its products but which company can help them treat their patients better. Who is conducting research into new chemical entities in their area, who has the ability to conduct new and exciting clinical studies, who can provide them with new up-to-the-minute information sources, and who can help them create better patient services?

Types of Franchises

Generally speaking, two types of franchises are discussed within the pharmaceutical world: complementary franchises and therapy franchises. A third exists, but is less well identified.

Complementary Franchises

Complementary franchises would include product brands that work together (e.g., both being prescribed to the patient at the same time either for different aspects of the disease state or as adjunctive therapy). A relevant example of this would be a cytotoxic oncology agent being prescribed with an antiemetic and a recombinant G-CSF for the reduction in duration of the associated neutropenia. A single company that creates a complementary franchise via a combination of its own pipeline and in-licensing is in a stronger position to leverage product brands with their prescribing physicians. Advantages can accrue, in particular, in low-competitive-intensity categories of products where the strength of the franchise can make the difference when the physician decides that they will support the company that supports them. BMS has successfully created a complementary franchise in oncology, as shown in Table 5.1.

Therapy Franchises

Therapy franchises would include product brands that all treat one indication, e.g., separate product brands that treat elevated blood pressure or high cholesterol from one company. Looking at the hypertension indication, a company with a number of different classes of product within their portfolio will act with more credibility than a company with a single product brand. Merck has two ACE inhibitors in Renitec/Innovace (enalapril), Cozaar (losartan), and Hyzaar (losartan and hydrochlorothiazide), as well as various diuretics around the

TABLE 5.1. Bristol-Myers Squibb in oncology.

Bristol-Myers Squibb complementary franchise
Numerous major products, indications, studies
The best-informed sales force
Diverse business development success
Huge KOL commitment
Owned, until recently, its own direct distribution network in the United States via Oncology Therapeutics Network (OTN)
Prepared to take bigger risks; e.g., Imclone collaboration

world, including Amizide (hydrochlorothiazide and amiloride). In this way, Merck tries to dominate the area and the intellectual direction of treatment for that indication through its therapy franchise (see Table 5.2).

In addition to complimentary and therapy franchises, which are relatively well known, a new type of franchise is emerging. It is termed the product origin franchise, which aims to try and simplify a multinationals structure to its varied customer groups.

Product Origin Franchises

Product origin franchises describe how a company may want to position different aspects of its business in different ways. Therefore, the Sandoz division of Novartis could be termed a generic franchise as it has created the single biggest global generics business through acquisition in the early postmillennium years. Novartis is the mother corporate brand; it markets patent-protected product brands too. The same argument could be applied to biotech subsidiaries that have separate branding from the master company brand. For example, Ortho Biotech and Centocor are part of the diverse multinational Johnson & Johnson (J&J).

The driving concept behind successful pharmaceutical franchise brand creation is that it is customer driven. At its strongest, this will be achievable when there is a coherent specialist customer group existing on a global level (e.g., oncology physicians, and respiratory or gastrointestinal experts). These relatively small groups of physicians (often 10,000 or less globally) are welded together by their similar desire for highly specific clinical and scientific data to improve their

TABLE 5.2. Merck in hypertension.

A therapy franchise

    Senior management understanding and commitment

    Willingness and confidence to invest in "landmark studies" that may well aid class acceptance

    Willingness to take risks with NCEs

    Being ahead of the game as far as trial outcome measures

    First on the list for potential in-licensing

---

day-to-day decision making. They are the kind of physician group who can reject mass promotional resources and huge numbers of poorly trained representative calls, instead preferring more thoughtful and less aggressive promotion from small and highly trained sales forces who live and breathe their area of expertise.

Any organization wanting to create a franchise brand has to build up a complex understanding of the needs of the customer segments and the groups/suppliers surrounding it. Franchises work best when there is clear cross-functional alignment within the business unit or division and the franchise strategy enjoys long-term commitment from senior management within the context of the overall company's portfolio. The coherency of the customer group is important in that, for instance, a CNS franchise is probably too wide a therapy area grouping to make sense; not only are there very different target physicians involved (ranging from neurologists to psychiatrists and PCPs), but the range of conditions covered under the general term is vast. Therefore, psychiatry or neurology could be realistic franchises on their own, but CNS would not resonate as strongly with the target customers as the breadth is too wide to be relevant. In the same vein, a cardiovascular franchise is difficult to establish, as the breadth of the product line and the different physician customer groups makes establishing a real franchise difficult. Therefore, BMS does have a franchise in oncology, likewise GSK in the treatment of AIDS, while MSD is more of a "cardiovascular house" that therefore finds it harder to leverage its assets into the wider cardiovascular area and achieve a true franchise brand position outside hypertension.

ADVANTAGES OF OWNING A REAL FRANCHISE

The customer focus that comes as part of franchise building effectively builds barriers to competitor market entry. Those barriers may be built from:

- a relationship perspective;
- a data perspective;
- a congress perspective; or
- a market-shaping perspective.

Particularly in smaller niche franchises, where there is a powerful but small group of global key opinion leaders, the incumbent franchise owner will be in a good position to shape the market and, importantly, be aware of competitor moves at the earliest stage. The vast majority of key clinical studies, their objectives, and protocol designs will be known within the clearly defined global physician community that the franchise owner has spent years cultivating. This power of market information and access to key physicians gives time for the franchise player to react to the situation. Knowing the likely profile of the competitor product brand well in advance of launch allows counter-profiling studies or merely spoiling studies where all the major clinical research studies are tied up with work for the dominant player. From a data perspective, analyses of previous studies can be conducted once there is insight into how the new entrant will differentiate and position itself, with a view to countering or nullifying the launch claims. Relevant publications and presentations can be made at the major global congresses in advance of launch to position the new competitor narrowly and, at its most extreme, a franchise owner can try to move the market away from the strengths of the entrant via market education.

The added credibility and track record of success over a number of years helps when the franchise company wants to launch a new product offering; in reality, it allows gain in market share at a discount because a huge amount of the launch preparation work is taken care of in day-to-day business. The physician group trusts the brand offering, and the positioning is likely to be right given the level of customer insight within the organization.

Another advantage is that allocation of resources can be made across a wider range of products while the internal organization remains focused on priorities. Selling synergies accrue as more than one product brand can be promoted effectively by the sales force, and if the brands are complementary to the physician target group, then they increase their usefulness significantly. In those areas, where DTCa is cost effective to educate the patient population about the possibilities of treatment, again synergies can accrue if a well-conducted franchise brand stimulates response for a number of possible product brands via direct response, Internet activity, or by merely motivating the prospective patient to ask the treating physician about product brands from the company or division in question.

Being well known for its proven expertise and depth of knowledge in an area also makes the recruitment of skilled and committed individuals easier. This works again to the advantage of the company with regard to retention of talent, as it builds genuine relationships with the key physicians and patient association leaders, who often become part of its social as well as business life.

As a result of in-depth customer and therapeutic knowledge, it is easier for clinical and marketing groups to have a common vision of what is required in terms of new product development and also life cycle management for existing product brands.

On occasion, a franchise business unit intrinsically "knows where to go" and what the key actions are. In these cases, senior management, which probably has come through the ranks, is aligned not only with the customer but also the key members of the business unit. In-depth understanding from representative to boardroom allows rapid decision making, measured risk taking, and speedy alignment in a way that can drive the franchise forward faster. In the late 1980s and early 1990s, the "Bristol oncology division" within BMS displayed this ability. They coordinated good internal pipeline with excellent business development activity to build a franchise that was comfortably the dominant player in cytotoxic therapy on a global scale.

The only real danger associated with the franchise concept is particularly related to the R&D pipeline.[2] Galenic development demands from the franchise may make management of the overall pipeline portfolio more difficult, as it will require considerable resource allocation to fit the franchise customer requirements, thus causing stress on the overall R&D and business development budgets. From a people perspective, developing in-depth knowledge in one therapeutic area doesn't mirror modern-day HR management within the industry. The pharmaceutical career that requires a job function move every two years is distressingly common among the high flyers, and, as a result, they gain incredibly varied experience in a highly complex industry but no real customer depth as their therapy areas and customers change so regularly. It is not uncommon for KOLs to remark that they no longer know who's in charge in a top-ten company because everyone moves so frequently and the vital top-level relationships are destroyed by the next inevitable job move.

Figure 5.1 shows that there are ample strategic options to expand the brand sales within a franchise and, as a result, its associated credibility and impact with customers. They range from traditional galenical development to OTC switches to value-added generics, and provide focus to the business development and alliance activities.

Although probably not strong enough to be classified as a true franchise yet, UCB S.A. has been able to compete against the biggest companies in the world within a specific niche—that of epilepsy. Although there are many sufferers globally, the treatment of epilepsy is not a major economic issue (as many sufferers don't contribute econo-

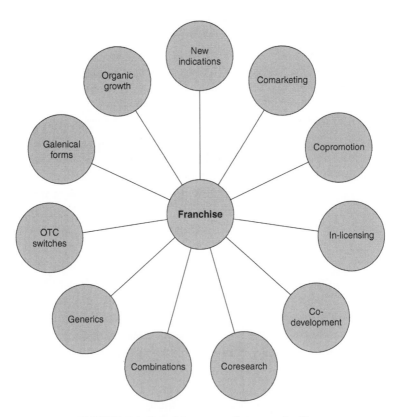

FIGURE 5.1. Franchise expansion opportunities.

mically due to the nature of the disease), and it isn't a "sexy disease" with "celebrity" personalities championing the issues and needs of the patients (although there are a number of very famous sufferers, they are paranoid that their neurological problem will be seen by the wider world as a mental illness and as such refuse to come forward to publicize the plight of patients). As a consequence, the epilepsy community worldwide is quite small, both the physicians and patient support groups are well known to each other, and they share many of the same objectives.

UCB S.A. launched a new second-generation antiepileptic drug (AED) brand called Keppra (levetiracetam) in the U.S. market in April 2000 with interesting data that showed a higher rate of seizure freedom in very refractory patients treated during the registration process. Within three years, Keppra managed to become the number one prescribed second-generation AED in the United States. This was despite being up against significant product brands such as Lamictal (lamotrigine) from GSK, Neurontin (gabapentin) from Pfizer, and Topamax (topiramate) from J&J, which all enjoyed more promotional spending and manpower. The way this was achieved was by leveraging small size, customer focus, a high-quality sales force, and, most significantly, first-class clinical data in a way that quickly brought broad support from the physician base.

Keppra's success has been repeated in all markets worldwide, with the exception of Japan, where none of the major new second-generation AEDs enjoy approval yet. Part of this success can be attributed to moving the market expectations. In the past, success was measured as the number of patients who achieved a 50 percent reduction in seizures, and education has moved this to the concept of seizure freedom, something that dramatically improves the prospects for the patient concerned. At only approx $2 billion in total, the epilepsy market is not large, and there are bigger markets to target with AEDs that include neuropathic pain and migraine. UCB S.A. is now busy trying to create a true franchise position, by following up their initial success with two new AEDs already in clinical development and innovative dosage forms for Keppra. To supplement this position, a focus within business development should bring other offerings for the neurology physician population and strengthen the franchise from being currently therapeutic to being complementary and therapeutic in the future. This could turn out to be a classic case of one good product brand speeding the achievement of a franchise brand that can be leveraged over an extended period.

FRANCHISE CRITERIA

Few real franchises currently exist despite a huge amount of discussion about the subject within the industry. To help in defining which "claimed" franchises are likely to be leveraging their own brand (and the product brands contained within them), the following list outlines some criteria and measures:

- Product brand status
  - Number of products within the franchise
  - Whether their galenical development suits the chosen franchise customer base
  - Breadth of indications for the targeted specialty
  - Longer demonstrable protection, once generic erosion of product brands starts
- Global geographical presence (otherwise it's a regional or local franchise)
- Franchise sales ranking, relative market share, and relative success of new launches
- Achievement of higher effective share of voice (SOV) despite having fewer reps
  - High sales force reputation with the customer base—quality and focus count
- Easy access to global key opinion leaders
  - Relationship building over decades counts too
- Being thought of as being first on the list of candidate companies when an in-licensing opportunity comes up within the franchise
  - Number of in-licensed product brands
  - Number of business development deals
  - R&D alliances
- Significant clinical development investment in comparison to franchise competitors
  - New products being developed

—The initiation of landmark studies

—Continued long-term depth and focus

—Scientifically interesting studies conducted as well as registration studies

• The presence of senior management (ideally the CEO and
COO), who also understand the customer group.

FRANCHISES—AN UNDERUSED OPPORTUNITY?

Despite the discussion, and the understanding of the many advantages that could accrue, not many true franchise brands currently exist
within the pharmaceutical world. There are multiple reasons for this,
starting with the obvious that there is clear confusion about branding
within the industry and its potential value, and, more significantly,
franchise building requires long-term senior management focus not
only over a five-year period, but over decades. This long-term strategic
commitment is required in R&D and business development, as well as
in the commercial arms of the organization (which is the easier part),
and tends to be associated with customer knowledge stretching from
representatives to the boardroom. To really compete effectively within
a particular franchise also requires a change in mind-set for the established patented product companies. Numerous "value-added generic"
opportunities exist within niches and traditionally run companies
aren't set up to develop both patented NCEs and new formulations of
older product brands (for which they may have to pay royalties). It
could be argued that this type of approach allows a company to gain
revenue at the fringes of its business model, but, in reality, an organization is going to find it extremely difficult to prioritize precious development resources for a product they don't "own." As a result, franchise
brands tend to remain unfulfilled as far as the influence they could exert is concerned. A dramatic change would have to happen.

WHO SHOULD BE BUILDING FRANCHISE BRANDS?

The easiest answer to this is to say mid-cap pharmaceutical companies and smaller, single-product or biotech-focused groups. For
big pharmaceutical companies, it is probably easiest to focus on the
corporate brand and try to optimize the obvious benefits branding

could produce. The problem for the biggest organizational structures is that they are not set up to create depth of customer insight, and unless some bold structural steps were to be taken, akin to the centers of excellence R&D approach of GSK, then the stress created across the organization by franchise management in all its guises would be too much.

That, therefore, leaves the mid-sized players, who either can't build the size required to compete effectively in mass markets, or don't want to. Creating strategic franchise brands is an excellent way of appearing bigger than your size competitively, punching above your weight, and gaining the many benefits from organizational simplicity. A good example is Lundbeck, who have a clear positioning within the CNS therapy area and will no doubt develop their position over time to dominating a number of franchises within that huge area.

For the smallest of entities, such as fledgling biotech or one-product brand companies, they are probably best sticking to a product brand approach until their size and success warrants a more sophisticated approach.

The mechanics of the branding process may be:

- Use of the company name
- Use of a subsidiary or operating unit
- Use of a separate franchise brand name

CONCLUSIONS

The terminology "pharmaceutical franchise" is used to signal a group (or range) of product brands that target a similar indication or physician segment (from a single company) with the objective of leveraging the brand equity inherent from previous (or future) product brand offerings.

There are three types of pharmaceutical franchises: complementary, therapy, and product origin franchises. The driving concept behind successful pharmaceutical franchise brand creation is that it is customer driven. At its strongest, this will be achievable when there is a coherent specialist customer group existing on a global level.

Any organization wanting to create a franchise brand has to build up a complex understanding of the needs of the customer segments and the groups/suppliers surrounding it. Franchises work best when there is clear cross-functional alignment within the business unit or division and the franchise strategy enjoys long-term commitment from senior management within the context of the overall company's portfolio.

The customer focus that comes as part of franchise building effectively builds barriers to competitor market entry. Those barriers may be built from a relationship perspective, a data perspective, a congress and promotion perspective; and a market-shaping perspective.

Significant franchise expansion opportunities exist from both internal and external sources. Organic growth, new indications, and OTC switches expand the franchise from an internal focus, while comarketing, copromotion, and in-licensing can expand it from an external perspective.

Despite the discussion and the understanding of the many advantages that could accrue, not many true franchise brands currently exist within the pharmaceutical world. There are multiple reasons for this, starting with the obvious that there is clear confusion about branding within the industry and its potential value. More significantly, franchise building requires long-term senior management focus, not only over a five-year period, but over decades.

Those companies that should build franchises are the mid-cap companies and smaller single-product or biotech-focused groups. For big pharma, where customer focus and depth of insight are more difficult to retain, it's probably easiest to focus on the corporate brand, as well as thinking about the obvious benefits that branding could create.

Chapter 6

Developing Brands

There are many books out there exhorting one particular process or another in development of brands. The process is perhaps of secondary importance. What matters is that at least three intellectual steps are completed.

First, there is a need to develop the brand's strategy by identifying what the brand identity will be, how the brand will be differentiated from competitive brands, and what the right target segment is. This first phase has to be based on in-depth understanding of the market, the consumers, the competitors, the environment, and the resources of the company to find a brand identity that is unique in the market and based not only on a functional clinical benefit but something both tangible and intangible that appeals to more than the merely rational side of human nature.

Second, it is necessary to develop a marketing program that is fully coherent with the brand's identity. This will be done by developing a program where every element of the mix will help communicate this identity to the target customers, e.g., not just the launch image and its current indications and pricing, but also future indications, formulations, delivery systems, etc.

Third, regular monitoring of the brand image is vital; the way the brand is perceived by customers is required to verify if there are differences between the identity that the company wants to convey and the brand's image. If a difference is identified, the company has to adapt its sales and marketing programs and better communicate their chosen identity to the target consumers. The process is summarized in Figure 6.1.

Pharmaceuticals—Where's the Brand Logic?
Published by The Haworth Press, Inc., 2007. All rights reserved.
doi:10.1300/5836_06

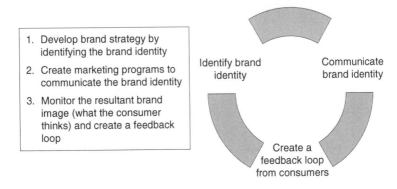

1. Develop brand strategy by identifying the brand identity
2. Create marketing programs to communicate the brand identity
3. Monitor the resultant brand image (what the consumer thinks) and create a feedback loop

Identify brand identity

Communicate brand identity

Create a feedback loop from consumers

FIGURE 6.1. Brand development in three steps.

Firms that rely on building strong brands as a major competitive advantage, such as the FMCG companies, will integrate the brand strategy and thinking very early in the development process. All departments will work together towards the brand development goal.

The pharmaceutical product brand represents, in the consumer's mind, a set of tangible and intangible benefits. They not only deliver a certain efficacy (tangible) but also offer additional values such as hope, relief, or trust (intangible). Pharmaceutical brands, therefore, have created strong associations in doctors' and patients' minds—pharmaceutical products have in fact all the elements that make up brands.[1] As in FMCG, there are subtleties represented across the different therapeutic areas (categories) within pharmaceuticals. A new disease area with high unmet medical need will be driven almost exclusively by clinical data, while a more mature and highly competitive market, concerning a curable disease without mortality, will rely more on sophisticated sales and marketing, of which branding is an essential element.

Pharmaceutical companies have not worked proactively in identifying a brand strategy that includes brand identity for their product brands. The brand image has often in fact been created by itself, without much control of the company in question. Arguably, the story of Prozac is a good example; the consumer adoption of Prozac as an icon for the 1990s produced a rapid and uncontrollable phenomenon. Lilly was faced with an issues management challenge rather than a brand management challenge. As a result, Lilly did not clearly define the brand—leading to significant variation in brand identity and therefore also brand image across markets. There was use of different arguments and clinical data sets with physicians across different markets in an uncoordinated fashion. The brand got out of control, becoming an icon for a certain way of life in the United States while alienating physicians in other countries (e.g., Belgium, who refused to prescribe the brand heavily as a direct consequence of patient demands).

Brands exist in both the FMCG industry and the pharmaceutical industry. The key difference between both industries lies in the fact that the FMCG industry is managing its brands with extreme care and investing most of its resources in developing long-term brands. Brands are the key assets of the company that will translate into financial value. The pharmaceutical industry, on the other hand, has not yet understood and integrated learning about the competitive advantage that brands could represent.

GLOBAL BRANDING

In the nonpharmaceutical world, the move toward global brands, rather than local ones, has been rapid over the last twenty years. The phenomenon was signaled by a famous article written by Ted Levitt in a Harvard Business School paper in 1983,[2] which argued that multinationals should "exploit the economies of simplicity." Adoption of the central arguments within it have led to a different consumer world, one where global brands appear to be the only things worth spending money on from a corporate point of view. Multinationals have drastically trimmed their portfolios of promoted brands, selling local brands to medium-sized or small local companies and focusing their promotion on building fewer global brands. Companies such as L'Oreal are now reaping the benefits from those acts of concentration as do Procter & Gamble (P&G) and Unilever. The central tenet to Levitt's argument was that multinationals could grow by selling the same product all over the world. In doing so, they could benefit from economies of scale, of having standardized product platforms and by expanding into new markets as a means of growth. This standardization allowed manufacturing and purchasing to benefit from economies of scale, making rapid geographical expansion possible and financially rewarding. The pressure to globalize brands continues to be significant. This resulted from the need to find new competitive advantage and pressure from the financial community.[3]

Local brands exist in one country or in a limited geographical area. International brands adopt some elements of the global strategy but execute locally, while global brands adopt the same strategy and marketing mix in all markets.

Global branding consists of offering a brand that has standardized a maximum number of elements of its strategy and marketing mix to ideally offer one near identical product brand to every international market. The principle to follow is, therefore, to look at what is common between markets and minimize or forget the differences between them.

Some leading consumer authors consider marketing globalization irreversible due to the important economies of scale permitted the emergence of global consumer segments and the rapid diffusion of technology.[4] Others believe, on the contrary, that global marketing represents a risk because differences of culture and consumer habits would remain between markets.[5,6] Today, global marketing has been adopted by the majority of FMCG firms. The question is not anymore whether to globalize brands, but rather to assess how to do it successfully. Globalizing at any cost doesn't make sense, and many of the big proponents of this strategy have, in fact, been pragmatic in their implementation. The same global product doesn't have to be the same global brand; for instance, Ariel in Europe is Tide in the United States. It is important to note that the creation of global brands has been more driven by cost considerations than consumer satisfaction.[7,8] According to Kapferer,[9]

> The decision to maintain or not a local brand in the portfolio should not be taken within a vacuum. A portfolio is much more than the sum of its brands: it is a system, an integrated whole, where each brand plays a specific role vis-à-vis key objectives such as market coverage, building category penetration, financing the introduction of high-end brands, and matching distributors' demands for differentiation.

Killing off a highly profitable local brand, where a global brand is not well known, or particularly suited to the culture of the local market, doesn't make sense financially, and, therefore, care has to be taken in the decision making. Another side effect was that in many parts of the world, local consumers had trouble relating to simple global propositions, and as such the term "international" and "glocal" started to be used, i.e., maximize product features, distribution, and selling techniques globally (to achieve those economies of scale), but adapt to local tastes in their branding.

While converging consumer segments makes sense (i.e., similar customers in different countries had more in common than different consumers from the same country), it is also argued that global brands create an aura of quality around themselves. This quality perception allied to being on the leading edge in turn creates a differential advantage in itself. An additional factor that has driven globalization is that free trade has been improving, and in theory still is, as an added aid to adopting this geographical expansion strategy. The downside for brand managers, however, is that global distributors are starting to demand global deals as they exert their own purchasing power—a phenomenon personified by Wal-Mart nowadays.

ARE THE DRIVING FORCES THE SAME IN PHARMACEUTICALS?

Pharmaceuticals have followed a similar path over the last thirty years as consumer goods. As successful multinational pharmaceutical companies have had to continue to look for sources of new growth, they have extended their geographic boundaries. Only fifteen years ago, many of the top-ten players were not represented in Japan, while those Japanese companies with presence in Europe and the United States were even fewer. Nowadays, we are rapidly approaching the stage where the same global companies compete in all markets on approximately the same scale. There are, however, relatively few mid-cap companies and biotechs that can genuinely call themselves global—but that will come in time with the targeting of new products to more specialized patient and physician populations. As in the consumer goods industry, having exactly the same brand name in every market is not mandatory, the name itself should not make or break a product. The GSK SSRI paroxetine is marketed under the brand names Seroxat, Deroxat, and Paxil (among others), while even the world's biggest selling drug in 2004, Lipitor (atorvastatin), is also sold as Sortis in Germany, Tahor in France, and Citalor in Brazil. Some local brands, especially in Japan, still thrive within individual markets but are becoming less significant as the industry changes. What is clear is that local brands are not going to drive growth across the world pharmaceutical markets during the next decade.

A number of synergies or economies of scale are thought to come from global pharmaceutical branding and its worldwide implementation.[10] A single R&D positioning and consistent branding throughout the long clinical trials phase of drug development are difficult to argue with. Manufacturing less variations of the standard dosage form and concentration of production in fewer sites are also easy to accept. Similarly, harmonization of treatment algorithms around the world for a particular disease means that everyone talks in the "same medical or scientific language" and, therefore, assesses the data in a similar, consistent fashion. In addition, the simplification of key opinion leader communication, the way they interact with other physicians, and, later, their willingness to provide endorsement on a global scale are simplified by a global brand. Last, but not least, the preparation of OTC switches is able to benefit from cross-border advertising and awareness in an increasingly mobile world.

When looking at the increase in global communications and the convergence of customer segments, it becomes clear that the same phenomenon is happening in the pharmaceutical world too. Patients now access the Internet, watch global television networks, and even participate in global support networks (at one extreme), suggesting indeed that segmentation of patients as well as consumers is becoming more relevant across country rather than merely within. Physicians, our main target audience, travel more frequently, and they are more likely to be involved in global peer groups, especially if they are working in specialist disease areas. We may also assume that diseases are the same the world over, as a result of which the pressure to globalize will be even stronger. Some important regional differences do exist, such as the problem of malaria in Africa and Asia, but there is little variation when considering the top seven markets (the top seven being the United States, Japan, Germany, France, United Kingdom, Spain, and Italy). So, a number of the major arguments for the creation of global brands in the consumer world are fulfilled. In addition, it is a moot point whether or not global brands in pharmaceuticals have managed to create the positive "aura" that branded consumer goods appears to have, but it could be argued that the ever-increasing regulatory standards demanded nowadays (and global standardization across regulatory agencies) does create that reassurance for physicians and patients in the know. Free movement of goods impacts pharmaceuticals in a manner similar to the consumer world; it allows "grey" goods to be shipped from low-price markets to high-price markets (like Levi jeans). This phenomenon, called "parallel trade," is the same in pharmaceuticals, and is one of the major points that detractors of the global branding model make. Like consumer goods companies, pharmaceutical firms are starting to be under increasing pressure from the financial communities and shareholders to go global, in an increasingly cost-conscious industry sector.

There are, however, several aspects that don't yet hold true. Global pharmaceutical distributors don't exist as such, although many national companies are expanding across borders or entering into alliances. The pharmaceutical business model is not put under the same kind of margin pressure by global distribution companies wielding huge buying power, but instead we are restricted in virtually every market by the governments and insurance companies that foot the bill. But, perhaps the most controversial point, and therefore most surprising, is whether or not globalization in pharmaceuticals leads to economies of scale.

In theory, it is easy to imagine that this must be the case, e.g., clinical studies that can be used for multiple regulatory agencies must be cheaper than separate development programs. Single product brand name global promotion must save money locally when it comes to awareness building. Local sales, marketing, and public relations activities must be more efficient and cheaper if it is done once centrally rather than reinvented by every product manager or local agency everywhere. That's the theory but, at this stage, unfortunately, there doesn't appear to be evidence of major savings—so what is the takeout? It may be as simple as the fact that due to pharmaceuticals being an industry where cost containment is not the driving force for margin improvement, we just do it (or measure it) very badly! This happens at all levels, and an easy example can usually be found in global marketing. Most global teams will be able to tell stories of how local affiliates have spent huge amounts of money on trying to beat the

Consumer goods	Pharmaceuticals
• Geographical expansion necessary to drive sales growth	• Geographical expansion necessary to drive sales growth
	• Standardization of clinical trials, data, and KOL management
• Standardization of manufacturing	• Standardization of manufacturing
• Purchasing economies of scale	• Purchasing economies of scale?
• Consumer needs are the same	• Patient needs and their diseases are the same
• Customer segments converging	• Consumer and physician segments converging
• Creation of an aura of quality	• Creation of an aura of quality (possibly true)
• Free trade improving	• Free movement of goods increasing
• Global distributor demands	• Coordinated government and payer demands for data and lower prices
• Financial market and shareholder pressure	• Financial market and shareholder pressure

FIGURE 6.2. Drivers of global branding.

global campaign. In so doing, they have often recreated a campaign that is not only wasteful but also is not in line with the agreed brand values when remeasured against the original criteria. If the cost savings associated with global operation were a focus, we would find out some home truths concerning implementation.

In view of the FMCG experience, the trend will be similar in the pharmaceutical industry, and Figure 6.2 compares drivers of global branding across the two sectors. We should expect the development of many more global brands and the elimination of many local brands, even successful ones. The pressure to reduce costs will be as important as in the FMCG area. It will be the key to further increase industry profits, and financial analysts and shareholders will continue to ratchet up the pressure.

ATTITUDES TOWARD GLOBAL BRANDS

Not everyone loves a global brand. In fact, branding, and American branding in particular (seen as cultural imperialism), has been under siege for some time now. Many multinationals are not seen in a good light because of their behaviors either locally or internationally, e.g., Nike, due to its targeting of lower-income youth in the United States or due to its use of child labor and low wages in the sweatshops of the Far East. Another example is Shell, following the proposed destruction of an oil platform in the North Sea (off the United Kingdom) or its exploitation and destruction of the environment in Nigeria. The attitude is best described and, as a result, best understood by reading Naomi Klein's best seller, *No Logo*.[11]

However, research carried out involving 3,300 consumers in forty-one countries by Holt, Quelch, and Taylor[12] showed there were far more important parameters for global brands than whether or not they were American; in fact, it simply didn't seem to matter to consumers whether the brands were American. The authors outline three dimensions of a global brand, which explain roughly 64 percent of the variation in brand preferences in twelve countries worldwide. These parameters were "quality signal," "global myth," and "social responsibility." Quality signal relates to the battles that consumers see between multinational companies when it comes to quality, and they like it. As far as global myth is concerned, this is about consumers using global brands to become part of a global group that they identify with. Social responsibility relates to the fact that consumers don't expect local companies to solve the issues of global warming, for instance, but, they do expect the global players like Shell or Exxon to do this.

According to the research, four global consumer segments were identified, global citizens (45 percent), global dreamers (23 percent), antiglobals (13 percent), and global agnostics (19 percent). Global citizens rely on a global brand as a sign of quality and expect the multinational to act with social responsibility. Global dreamers are less discerning and readily accept global brands and as a consequence the myths they are selling. Antiglobals are cynical about improved quality from global brands and if given a choice would not buy a global brand. Global agnostics don't base their decision making on whether a brand is global or not and only care about whether it satisfies their needs or not. (See Figure 6.3.)

The challenge for the future appears to be competing effectively against other global brands (price, performance, imagery, etc.), while managing the brands as global symbols. Consumers appear to be either

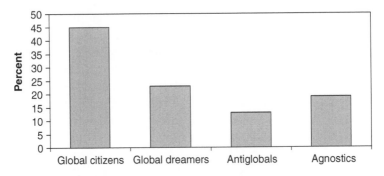

FIGURE 6.3. Global consumer segments (expressed as percent of total). (*Source:* Adapted from Douglas B. Holt, John A. Quelch, and Earl L. Taylor "How global brands compete," *Harvard Business Review,* September 2004, pp. 68-75.)

in awe of multinationals or greatly disturbed by them. As a result, firms need to manage their image within an unstable global culture through credible myths and credible actions. A famous example was British Petroleum's (BP) attempt to identify itself with a clean future environment, a myth that was easily overturned by activists, due to the credibility gap between the myth and its well-documented and publicized actions to the contrary.

GLOBAL BRANDS IN PHARMACEUTICALS

To date, this type of research has not been completed (or published?) either for physicians or patients. Having said that, physicians and patients are consumers, and, therefore, it could be assumed that the parameters established in consumer research may well be applicable to the pharmaceutical industry. So, both could be attracted by the amount of effort big companies put into competing for the "quality" position in the different drug categories (this is most likely represented by the number, size, and design quality of the clinical trials done by the company in support). Equally, a certain percent will be drawn by the myth around the drug brand—wanting to be part of that global group of users or takers of the drug brand (e.g., Viagra). Without doubt, social responsibility is increasingly coming into the pharmaceutical headlines. In the past, this would perhaps have been interpreted as a positive factor, but nowadays barely a week goes by without reports of another antitrust or similar marketing violation committed by the top players within the industry. A whole chapter of this book is devoted to the notion that the industry has been, and will be in the future, under constant attack from external stakeholders, and dealing with that issue is not going to be easy. The pharmaceutical industry has its own issues, which mirror the problems that Microsoft has had in the courts over anticompetitive activity, Shell has experienced over the environment, McDonald's has had to face along with Nike, and a host of other high-profile global brand names.

If the challenge of the future for consumer companies is "competing effectively against other global brands while managing the brands as global symbols," then the pharmaceutical industry has the same problem. Continuing the theme of myths from the consumer section above, one obvious example has been given by Pfizer. In the second half of 2004, Pfizer, in the United States, launched a new scheme to give uninsured patients a discount on their medicines. The discounts get bigger depending on insurance coverage and ability to pay—something that surely must be a good thing? The problem appears to be that calling the scheme "Pfizer Pfriends" immediately hit a cord with anti-industry commentators such as Roy Lilley, who found it patronizing for lower income groups[13]—another example of a myth being easily overturned due to a credibility gap and something the industry has to become increasingly sensitive to.

DIRECT-TO-CONSUMER ADVERTISING (DTCA)

Few pharmaceuticals manage to become adopted into global culture, enter the language, and become a metaphor for the health of society and the strength of prescription medications. The end of the 1990s, however, saw a rise in this phenomenon with Prozac and Viagra entering global consciousness and achieving global brand recognition, in part because of changes in United States legislation allowing DTCa advertising from 1997 onward. DTCa was heralded at the time as a new dawn for the industry and a driver of future growth. With it came a rush of enthusiasm for consumer marketing practice and expertise. So far, the results have tended to be disappointing and haven't yet looked like they are going to revolutionize the industry. It has merely become another part of the marketing mix. According to research by Health Products Research in 2004,[14] 80 percent of a product's share change can be tracked to detailing and sampling, with DTCa achieving only a paltry 10 percent—so the impact of DTCa hasn't lived up to the hype.

The traditional focus for pharmaceutical promotion, and therefore branding, had been up to this point the prescribing physician. This was particularly true two decades ago when each physician had the "god-given right" to prescribe exactly as he or she wished, in the best interests of the patient in question. Gradually, this position of absolute authority has been eroded, and now the previously unseen hand of the payer is becoming ever more clearly evident. The physician is losing control, while the payer and patient have increased their respective power positions, and this trend is likely to continue going forward. The need for cost control will become even more critical to governments, and insurance entities, and the patient influence will, in addition, increase as he or she shoulders more of the financial cost of the prescription decision in question. Payers need to understand the impact of a particular medicine and assess the value of the product brand within the context of their budgets and administrative processes. They need to understand which patients should receive the product and which should not, what benefits a patient will accrue through taking it, and what the budget implications are going to be. Key questions such as whether usage will reduce disability, provide earlier release from in-patient care, or improve longevity and wellness needs to be answered. In most countries, health care provision is a political process, and, therefore, payers also need to be able to understand what would be the social and resultant political outcomes of using the said medicine. DTCa and branding don't fit in with the information needs of the payers, as their role is cost management to overcome excessive branded product usage. Despite this, they remain consumers just like everybody else.

A better understanding of the impact of this new media of DTCa has been reached in the United States over the last decade, and it has become an increasingly important part of the mix for a relatively small number of big brands. IMS Health data, reported in *Pharmaceutical Executive,* showed that nearly $3.5 billion was spent on DTCa in the year up to March 2004, with the top-ten brands accounting for a third of all spending. This is in comparison to detailing where the top-ten products only accounted for 15 percent of the total $10 billion spent. So, DTCa appears to work for fewer brands than detailing, if the spending is a good reflection of return on investment (ROI). The reach of that pharmaceutical advertising is increasing at the same time, moving from the original simple print media in consumer titles to mass television advertising. Even more targeted sports sponsorship is appearing, which looks to capitalize, for instance, on the male population's enthusiasm for sports such as American football (even though sports sponsorship is notoriously difficult to measure in its effectiveness). Cialis (Lilly) has sponsored the PGA Tour (golf), while Viagra (Pfizer) has been involved with Major League Baseball, and Levitra (GSK/Bayer), the NFL. The rationale for GSK and Bayer's sponsorship of the NFL (comarketers of erectile dysfunction (ED) competitor Levitra) is presumably that the brand associations of the NFL (speed, power, maleness) are close to the desired brand values of its ED product.

Both branded and nonbranded (disease awareness) advertising takes place, and celebrity endorsement is now a standard part of advertising tactics, with the erectile dysfunction brands leading the way, associating famous sporting faces with their product brands. These kinds of activities could be thought of as cobranding activities, and probably give enough space between the pharmaceutical product brand and the event it sponsors, to provide distance in the event of unfortunate incidents on either side. But DTCa remains a relatively blunt, widespread media, which wastes a huge amount of it's spend on target audiences that have no interest in the area concerned. There are no big-selling magazines devoted to fungal nail infection or the mentally disturbed, in stark contrast to consumer areas where there are many car magazines, yacht magazines, women's interest periodicals, etc. Generally speaking, DTCa advertising in the United States tends to benefit the whole class or category that is being advertised, and does not generally affect the share of the constituent competitors, given all other things being equal.

As yet, DTCa is banned pretty much everywhere else in the world (limited DTCa is allowed in Canada and in New Zealand, but there are moves to ban it in New Zealand in the future). This is probably more to do with payer concerns, about the costs it may create through patients specifically requesting a brand name medicine, than concerns about whether the consumer is capable of understanding the communication or not. DTCa and Internet branding strategies, therefore, depend on where you are geographically to some extent. If the full gamut of DTCa tactics is available in your particular market, then it comes down to choosing the most effective mix of components you can afford to reach and influence consumers (and their physicians and payers). Elsewhere, there will be major restrictions on what activities can be undertaken, and caution needs to be exercised to ensure the law is complied with.

THE INTERNET

The major advantage of the Internet is that it is pretty much a global phenomenon, and this, along with an explosion of medical information providers, has created an unparalleled opportunity for the patient to interact with medical information and gain knowledge about their medicines—only a decade ago, this would have been impossible. Physicians are also Internet users and, finding their time constrained by ever-increasing demands, are also turning to the Internet as a means to search for information. So far, the industry has responded by creating company Web sites that physicians and patients can access, giving product-specific, as well as disease background information. In addition, there have been successes with experimentation in e-detailing (product and CME), as a means of trying to improve person-to-person efficiency of promotion—although those physicians who like to be contacted by the industry in this fashion are a small segment of the overall population. The facts are, however, that both patients and physicians prefer independent Web sites to those created by the industry, most likely related to the myths around the industry and what our commercial intentions are perceived to be. Industry Web sites appear to be a necessity, but in reality garner little traffic. Their existence appears to be a basic information requirement by physicians, and if the communication tool is absent, it causes irritation. Payback is therefore marginal, with no single success matching, for instance, the success of the old Merck manual, which is still sold around the world in quality book stores. In an attempt to try and improve the situation, in particular for corporate or disease area brands, a number of companies have decided to sponsor independent Web sites that cater to patients' and physicians' needs with the added authority of being independent.

DTCa remains controversial;[15] health campaigners and consumer groups contend that DTCa can be dangerous by promoting new medicines rather than the best. Criticism is real because the quality is generally low (poor accuracy despite being preapproved by the FDA) and that it places the emphasis on cure and not prevention. These concerns and the public debate around it have forced the FDA into a position of increasing their scrutiny of advertising in response, and this may well make the creation of truly useful DTCa even more difficult in the future. Despite this, studies do suggest that as far as the consumer is concerned, DTCa has value for the patient and the immediate members of the family of the patient.[16] The adverts seem to generate considerable interest in talking about health-related issues with the physician and play a societal role in educating patients about their medications.

The thing about these two media (traditional DTCa and the Internet) is that they change everything, but perversely everything remains the same. Irrespective of a patient's desire to be treated with a particular brand name medicine, the most significant deciders of what is eventually prescribed are physicians and the governments or insurance companies that pay the bill. The industry is left struggling with different customers with different information needs (and influences), and it's therefore not surprising that the branding model for pharmaceuticals remains complex, poorly understood, and, unfortunately, largely unresearched.

PHARMACEUTICAL BRANDING COMPLEXITY

The popular view is that pharmaceutical marketing is far too complex to be able to use relatively simple consumer techniques. The complex matrix of interactions that exists between four key stakeholders, physicians, pharmacists, payers, and patients is complicated further by the very nature of disease itself and social expectations. These stakeholders have to take into account their own individual biases, e.g., how and where they were trained, as well as a patient's disease history. Other facets of decision making include the patient's likely compliance with medications, what payer restrictions apply, possible commercial substitutions (in the pharmacy) and the impact of patients needing to copay, and therefore, their ability to pay.

At this stage, branding within the industry has not led to increased revenues, although many who work in the industry themselves would doubt this. Who hasn't seen or experienced global key opinion leaders wanting to be involved in clinical studies for new products. The company's reputation and track record of establishing a global product brand and the likely size of their development spend are an inherent part of the attraction for KOLs. This acts in concert with their potential fame if important findings emerge.

Branding hasn't been acknowledged as impacting on the achievable price on a worldwide basis, but our current thinking is too tactical. It is thought that, due to pricing restrictions and lack of good pharmacoeconomics, those pharmaceutical companies are prevented from using the strength of the product brand to achieve a certain price or raise it. Whether governments and insurance companies like it or not, coordinated franchise brand and product brand communication does impact on the likelihood of listing in formularies and the potential price point achievable. In the U.S. market, increases in price on an annual basis are not unusual, particularly where managed care doesn't see the product as a cost priority and competition is relatively weak. Therefore, advantages can accrue if the process is well managed and branding opportunities maximized.

It's also argued easily that physician loyalty is assuaged by formularies and, therefore, the impact of branding is minimal. To some extent, that is correct, being in or out of the formulary is a gatekeeper to an extent, but inclusion in the first place is dependant on many factors—even insurance companies and governments have to let the people think they are getting the best care available, and that kind of "feeling" isn't a clear-cut financial decision. The cheapest cancer patient is one without treatment, as in the bigger scheme of things their few months extra of average survival isn't a useful use of resources. Rationing, triage, call it what you like, is increasingly a public debate (especially in Europe), but it is one few payers would encourage. Even within a rigid formulary, brands, and how they are promoted make a difference, e.g., clever use of data to provide quality-of-life advantages for one brand over another will win the day. What we must not forget is that there is no standard patient—each individual reacts differently to a pharmacological agent and, therefore, there is always room for debate, something the industry does not stress in its communications. The role of the physicians may be changing, moving away from being in total control, but their opinion is still a major part of the final decision making process.

Perhaps one of the reasons why DTCa hasn't been that effective is that it is purely a U.S. phenomenon. This leaves the global multinational with the problem of how to deal with it and integrate it into normal marketing practice. The lack of success with DTCa could be caused by the relative isolation of implementation within the United States, but is it really that isolated? Motivated patient groups, or individuals, are now used to simple questions guarding Web sites such as "Are you a U.S. citizen?" They say yes and consult the data and use it locally. Naturally, from a global perspective, this stops any realistic measurement of none officially sanctioned DTCa, such as patient group sponsorship, but there is little doubt that it does impact far wider than the U.S. market. As medicines become more targeted, this will only increase with time.

As far as global branding is concerned, it is argued that lack of local adaptation means benefits are not created. These differences can be important; for instance, a psychiatrist in Spain will without doubt treat a patient differently from a psychiatrist in the United States (but it could be argued that this is due to very limited numbers of pharmacological products being available for use). Looking at a more technically data-driven physician group, such as neurologists, would however see a very clear global coordination across all core markets and in fact across most of the rest of the world too. The argument goes that making a local PM toe the line reduces local effectiveness within their culture. An easy riposte is that, unfortunately, product management is still not seen as a significant line function within the industry; it is more of a passing phase to sales management and general management rather than a discipline in itself. Therefore, PMs are often young and chosen for their personality type—for wanting to change the world! Hindsight and significant experience has led me to the conclusion that I've changed things that didn't need changing in the past, tactical branding guidelines drive rigor and proof that change locally is required.

What doesn't help the industry is that pharmaceutical marketing is too often nothing more than a means of feeding huge sales forces. Some of the misconceptions about branding and where it fits in are, therefore, deeply rooted in the structural fabric of the industry we work in and the career paths that people take. One of the central issues behind the use of the consumer branding model is that it has been largely applied to only one area in the complicated web that makes up pharmaceutical marketing. It is no good taking one tiny part of the communications that a product brand might have with its multiple customer base (physicians, pharmacists, payers, and patients), applying a few consumer branding concepts, and expecting the whole thing to work. The problem is that DTCa and the other parts of the branding communications mix have to be integrated so that it makes sense on a macro level. It is no good spending millions on advertising that says "we care," when many of the stakeholders see the company as distant, arrogant, at times aggressive, and often anonymous. Blaming the branding theorists as pharmaceutical failures, or saying our industry is different, are not the answers either. The answer is to start thinking about the whole picture, the bigger picture, of selling innovative medicines to physicians, pharmacists, payers, and patients. Watching pharmaceutical DTCa television adverts is an uninspiring experience; you could probably switch the brand names to those of any life companies and many would not notice the difference. How many of these adverts effectively communicate to the patient what the most effective use of the medicine would be, and build on an inherent patient need associated with the condition they may suffer from? At this stage, far too few.

For consumer companies, brands represent a way to create a significant competitive advantage that can translate into barriers of entry; as such, brands represent a way to differentiate products against competitors. Brands also generate consumer loyalty; they signal a certain level of quality that the consumers will find on a regular basis. This loyalty will therefore provide a predictable demand for the brand that will translate into consistent revenues for the company.

For consumers, brands can represent a relationship of trust and loyalty; they find a certain level of quality when they buy a certain brand. There is a tacit contract between the manufacturers and the consumers that the quality of the brand will remain constant, in line with "what the brand stands for." Brands can also represent symbolic values, allowing consumers to project their own self-image or aspiration for living a certain lifestyle. At the same time, they represent some kind of guarantee against certain risks such as poor safety.

Many industries have adopted a brand logic. The FMCG category has been leveraging brands for many years, but now services, durable goods, and other industrial companies are also leveraging brands as a key competitive advantage. Before 1991, Intel was not known at all by end-consumers. Since then, however, Intel has reached a very high level of awareness, created a quality image, and changed the consumer decision making process when buying a PC. It was ranked as the fifth strongest worldwide brand in terms of value in 2002, worth more than $30 billion.[17] This was achieved by creating a brand identity and communicating it to end-users via a campaign integrating the "intel inside" logo in most computer manufacturers' advertising to create lasting loyalty.

Why Can't these Advantages Apply to the Pharmaceutical Industry?

From a company point of view, pharmaceutical products need to be differentiated from their competitors, especially now that product performance differences are getting smaller as the competitive environment intensifies. It is also essential to create loyalty from doctors and patients as generics play an increasing role. Brand loyalty can generate sustained revenues that are important to ensure predictable cash flows for the firm. From a consumer point of view, doctors and patients are looking for a guarantee of quality and safety that a brand can bring. Successful brands also generate trust—a crucial aspect for pharmaceutical products that have an impact on human health.

There is without doubt a complicated interaction between the different aspects that make up pharmaceutical promotion, sales force interactions, and brands. Experience from other industries has taught us that a simple focus on product attributes does not maximize either the product sales potential or its associated value as far as shareholder value is concerned. Product differentiation is quickly challenged nowadays, but brand loyalty can be created through factors that are more difficult to challenge.

CONCLUSIONS

There are three vital steps in developing brands; first, to establish the brand strategy and identity; second, to communicate the identity via a comprehensive marketing program; and third, to regularly monitor what the consumer perceives to be the brand image.

Many of the forces that have driven consumer brands to become global are equally present in pharmaceuticals, but important differences do exist. Consumer attitudes to global brands vary among global citizens, global dreamers, antiglobals, and global agnostics. The same research isn't available for pharmaceutical customers, but many similarities may accrue.

DTCa and the Internet change everything but perversely everything remains the same. Irrespective of a patient's desire to be treated with a particular brand name medicine, the most significant deciders of what is eventually prescribed are physicians and the governments or insurance companies that pay the bill.

Brand loyalty can generate sustained revenues that are important to ensure predictable cash flows for the firm. From a consumer point of view, doctors and patients are looking for a guarantee of quality and safety that a brand can bring. Successful brands also generate trust—a crucial aspect for pharmaceutical products that have an impact on human health.

Experience from other industries has taught us that a simple focus on product attributes does not maximize either the product sales potential or its associated value as far as shareholder value is concerned. Product differentiation is quickly challenged nowadays, but brand loyalty can be created through factors that are more difficult to challenge.

Chapter 7

Product Brand Longevity

CONSUMER EXPERIENCE

One of the major differentiators between FMCG branding and pharmaceutical branding concerns the longevity of the brands over time. Pharmaceutical product brand life cycles are measured in years rather than decades or indeed centuries. As already covered in Chapter 1, there are a significant number of well-known consumer brands that have survived decades in good health. The accompanying Figure 7.1[1] shows the longevity of some consumer brands; 64 percent of the most known brands in the United States (in 1991) were fifty years old or more, with 10 percent being older than 100 years. This doesn't mean that new brands have not become successfully established in the interim period, but that long life spans are common in the consumer world.

The perennial youthfulness of many current-day consumer brands makes it clear that brands can be continually updated and renewed as customer attitudes change. Coca-Cola was born in 1887, Michelin in 1898, and Marlboro in 1937, but they remain modern, relevant, and ever youthful. The successful long-term brands have tended to stay true to themselves over time, appreciating that they are their own benchmark, avoiding overextension, therefore managing to maintain or grow market share, and, as a result, keep distributor own brands in check over many decades. In this way, they have maintained profitability even when under pressure from numerous competitors.

By comparison, in the pharmaceutical world, no brand has survived with sales intact over such an extended period; aspirin may have retained product volume but has not been able to maintain prof-

Pharmaceuticals—Where's the Brand Logic?
Published by The Haworth Press, Inc., 2007. All rights reserved.
doi:10.1300/5836_07

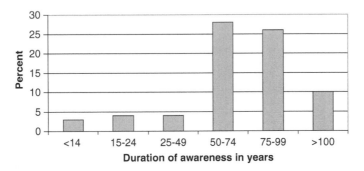

FIGURE 7.1. The Longevity of brands—the age of the most known brands in the United States. (*Source:* Adapted from Aaker, D.A. (1991) *Managing Brand Equity,* The Free Press. In turn adapted from Bogart and Lehman "What makes a brand name familiar.")

itability. Building drug brands, therefore, does not appear to protect long-term sales and profit in the same way as it does for other sectors. The major simplistic reason for this is that pharmaceutical product brands are limited to twenty years patent protection within the major world markets, with the potential for five years additional supplementary product certificate (SPC) coverage.

Generally speaking, successful long-term brands bring added value to the market through either tangible benefit such as improved product performance. For example, Gillette razors and their continual renewal and improvement, or intangible benefits, or Nike, where the attraction of ownership created within the young adult population has been equally idolized and demonized over the last two decades. The often quoted super brand represented by Coca-Cola has maintained with great success its chosen brand strategy of availability, affordability, and awareness over many decades. This is despite being under pressure from a better-tasting competitor in Pepsi (according to widely quoted consumer market research) and multiple other soft drinks brands.

Management of these successful long-term brands requires significant investment in market research and brand management. Customer needs and perceptions change over time, and, as a result, great effort is required to understand these changes and relate them to the current brand and the communication around the original brand, the key to the process being an in-depth understanding of the perceived

TABLE 7.1. Sustaining brands over time.

Significant research to understand how customer needs change over time.
Brand has to bring tangible and/or intangible benefits to the market place.
The perceived difference is nourished continually over time by management.
The brand stays true to itself and its range—overextension is avoided.

differentiating factors or points of difference. These factors, if successfully managed, can maintain a leading positioning in the minds of the consumer and sustain sales volume and profitability through continual nourishment of the brand. Table 7.1 sets out key factors for sustaining a brand over time.

DISTRIBUTOR OWN BRANDS AND GENERICS

When using insight gained from consumer branding work, it is a relatively easy intellectual jump to see the similarities between distributor own brands and pharmaceutical generics. Distributor own brands are copies of successful branded products sold usually at a discount (major or minor) to the original and targeting already established customer needs. According to commentators such as Kapferer, distributor own brands (the rough equivalent to generics in pharma) thrive in markets where there is high volume and where significant branded sales preexist. This is also true where the offer never changes (no evolution in the proposition over time, e.g., fizzy soft drinks), where brands are expensive, where customers perceive little risk in the purchase decision, where technology is cheap (and widely available), and where choice is often driven by the visual characteristics of the brand.[3] In addition, we can probably add to this list when customers have obtained and understood all the relevant information and where there are low switching costs, e.g., the choice of breakfast cereal on the one hand and the sharp contrast to choosing between computer operating systems, where changes can be technically very difficult to manage due to incompatibility of peripherals and other infrastructure. By default, this model therefore suggests that brands dominate when the factors are reversed, as portrayed in Figure 7.2.

In looking at the factors that favor brand domination, pharmaceutical product brands have a number of potentially favorable factors to

Generics dominate	Brands dominate
• High volume	• Low volume
• Significant brand sales	• Small sales
• Offer remains static	• Offer changes
• Brands expensive	• Brands cheap or underpriced
• Customers perceive little risk in choosing	• Customers perceive risk
• Technology cheap and widely available	• Technology expensive and not easily accessible
• Choice driven by visual characteristics	• Choice driven by multiple characteristics
• Buyer has all relevant information	• Buyer has limited information
• Low switching costs	• High switching costs

FIGURE 7.2. Factors in deciding generic or brand domination.

suggest that product brand domination is possible. The first two aspects are not highly relevant to making an FMCG versus pharmaceutical comparison, except to say that generic competition will target the high-volume product brand sales and low-volume, high-priced brands as a priority when patents expire. The area where the differences really start to emerge concern complexity. The offer for a pharmaceutical brand can change over time as new relevant trial or diagnostic information becomes available, and this then should be a pharmaceutical product brand advantage. Customers (physicians, patients, pharmacists, and payers) rightly perceive risk in the choice of pharmaceuticals, and the technology is generally expensive when new chemical entities are launched. In addition, the buyer (again, physician, patient, pharmacist, and payer) has limited knowledge of all factors affecting the patient in the vast majority of consultations, as there are always unknown factors in the prescribing decision process. Likewise, switching costs may be high, in that the side effect profile may not be fully known with an alternative, and this is exacerbated by the need for up-and-down titration of some medications.

In theory, major factors favor brand domination within pharmaceuticals. The major difficulty, which limits the development of branding in the pharmaceutical area, is that all of these positive factors are outweighed by the simplest of arguments—pharmaceuticals are limited by the patent life afforded by the authorities around the world. Once expired, rapid generic penetration eats up sales and profits within months, sometimes, but certainly a number of years. But often this simplistic argument hides many insights.

BRAND DECLINE

Brand decline or failure isn't purely a pharmaceutical phenomenon. Naturally, there are also many examples where consumer brands have failed to survive over the years even when their competitors have thrived. The major factors or causes of consumer brand decline are represented in Table 7.2.

The most easily identified falls from grace have occurred when an FMCG brand has been replaced by another technology (e.g., Betamax videotapes and eight-track tape players). After technology replacement, most of the other factors are associated with mistakes in the brand strategy—quality, communication, distribution, pricing, range mistakes, refusal to follow an enduring market trend, and, of course, withdrawal of investment. For example, a common mistake is the failure to maintain a quality that is crucial to the brand's long-term health. Squeezing out the last drop of profit through production savings can have dire consequences if the consumer becomes aware of

TABLE 7.2. Causes of brand decline.

Replacement technology
Brand strategy
Degradation in quality
Communication
Distribution mistakes
Pricing
Range mistakes
Refusal to follow a durable market change
Withdrawal of investment
Safety events

the dropping quality of the offering. It can then take many years to repair the damage. Many European and U.S. car brands suffered terribly as Japanese vehicles entered the global market and rapidly set new benchmarks for reliability; the most heavily affected companies included General Motors, Rover, and Lotus (United Kingdom).

Brand strategy and communication mistakes also commonly lead to brand death. Perhaps one of the greatest escapes was when Coca-Cola responded to Pepsi's taste pressure and changed its formula. Luckily, Coca-Cola quickly realized the mistake and took significant remedial action swiftly enough to retain its number-one position. Due to the importance of shelf space in consumer goods stores, it can be fatal to allow your brand to be trapped in an evolving or changing distribution agreement that gradually squeezes it out of the market. Pricing is without doubt one of the most sensitive areas in consumer goods marketing, with the now classic example of Marlboro, which, under pressure, cut its U.S. price for cigarettes and grew its market share over the ensuing months. On the flip side are many more examples of pricing decisions gone wrong. Range mistakes are another good example of where brand strategy can cause decline, e.g., overextending a brand into an area where the original name does not convey a strong enough differentiation. Bic makes great pens, pencils, and razors, but its image can't extend to perfume. A different kind of mistake is made when a durable change in the market is not immediately adopted. Taylor Made relinquished its number-one position in golf equipment when it failed to adopt bigger, more user-friendly heads on its drivers, an innovation originally pioneered by Callaway and now successfully copied by Taylor Made many years later. Disinvestment is probably the most easily understood factor; neglecting a brand will, over time, ensure that as competitor offerings improve, market share drops. Making the right disinvestment decisions can be difficult. The last, but perhaps not most frequent, mistake surrounds safety of the product offering. A few instances of car tire safety and other serious potential flaws have done immeasurable damage to major manufacturers' brands and have taken years to recover. In the consumer world, safety issues tend to lead to brand decline rather than product withdrawal and subsequent brand death.

Generally speaking, the vast majority of consumer factors are directly translatable to pharmaceutical brands in prescription and/or

TABLE 7.3. Examples of brand decline.

Cause of brand decline	FMCG	Pharmaceuticals
Replacement technology	Betamax tapes and eight-track cassette players	Part replacement of diuretics, beta-blockers, ACE inhibitors
Brand strategy		
Degradation in quality	U.S./E.U. cars versus Japanese. built quality	Pan Pharma generics in Australia
Communication	Coca Cola	Delay in Clarinex launch for Schering-Plough
		Claritin Rx to OTC switch
Distribution mistakes	Common in consumer goods-shelf position vital	Examples exist in OTC shelf position
Pricing	Marlboro cigarettes	Failure of brands to follow generic pricing initiatives
Range mistakes	Bic pens versus perfumes	Examples exist in OTC
Refusal to follow a durable market change	Taylor-made and big-headed golf drivers	Capoten b.d./t.d.s. dosing versus o.d. Innovace and Zestril
Withdrawal of investment	P&G portfolio review	U.S. market patent loss signals global divestment
Safety events	Motor manufacturers	Rezulin and Lotronex
Patent expiry	Legal process often to slow to impact	Prozac, Augmentin, Seroxat, and many others

OTC settings—see Table 7.3. Naturally, some factors are much less frequently seen (e.g., range extension mistakes or distribution mistakes). These types of mistakes are more common in the OTC area rather than the prescription side of the business.

Although virtually all factors of decline are relevant, the pharmaceutical industry tends to put most of its thinking into just two areas—those of patent term expiry and product safety. Many of the other factors are not managed in as clearly coherent fashion that would provide benefit.

Patents are naturally important in the consumer area too, but the situation tends to be exaggerated in pharmaceuticals where loss of exclusivity can lead to significant and rapid drops in the sales of the original brand. Prozac is an excellent example, which lost over 80 percent of its U.S. sales in just six weeks when it came off patent in 2002. This model of rapid loss has accelerated in the 1990s and early millennium compared to historical sales. During the 1980s, a product suffering patent loss could still expect to have 60 percent of its sales turnover twelve months later. In the 1990s, that figure dropped to 40 percent and, in certain cases, as in the case with Prozac, this has been further exceeded postmillennium.

Product safety is of paramount importance in pharmaceuticals where there are more examples of brand death (product withdrawal) due to adverse product safety than is common in the consumer area. Rezulin (troglitazone) provides a good example of how catastrophic a safety event can be. This antidiabetic was originally marketed and later withdrawn from the U.K. market by Glaxo in the mid 1990s due to potentially serious hepatic (liver) side effects. It was later launched in the U.S. market by the Parke-Davis division of Warner-Lambert only to befall the same fate and be withdrawn from the U.S. market on March 21, 2000. Brand death in pharmaceuticals inevitably leads to litigation, despite this threat Warner-Lambert was acquired by Pfizer (due to the significance of Lipitor (atorvastatin)). In its 2002 annual report, Pfizer said it was involved in 8,700 suits, including a class action suit by Blue Cross/Blue Shield of Louisiana and other health benefit plans, which were intending to recover $1.4 billion paid for Rezulin between February 1997 and April 2001.[5]

A second example that shows there is occasionally (but very rarely) a way back from pharmaceutical brand death concerns the irritable bowel disease brand Lotronex (alosetron hydrochloride) from Glaxo-SmithKline. Originally launched on February 9, 2000, the brand was withdrawn only a few months later on November 28, 2000, when GSK and the FDA were unable to agree on a risk management plan. This followed serious gastrointestinal events that had resulted in hospitalization, blood transfusion, and/or surgery, and some fatalities in the female population, in which it was indicated for use. However, on June 7, 2002, in a virtually unprecedented move, the FDA approved a Supplemental New Drug Application (sNDA) for Lotronex, which allowed reintroduction of the brand under restricted conditions of use. In reality, significant patient lobbying, of both the FDA and GSK, had taken place by those patients who had benefited from the product brand, including patient testimony in front of the FDA advisory committee, which recommended reintroduction.[6]

Patent expiry and product safety issues dominate the pharmaceutical industry thinking as causes of brand decline, but all the other consumer factors are relevant. Quality issues saw the decline of Pan Pharmaceuticals brands in Australia; their manufacturing license was suspended for six months in 2003 after the Australian Therapeutic Goods Administration (TGA) found a series of safety and quality breaches by the Sydney-based manufacturer. A recall was ordered of 219 products, and this was following a previous January 2003 action when an anti-travel-sickness brand, again from Pan Pharmaceuticals, called Travacalm, was found to have 0-700 percent variation in its active ingredient. Action was taken on Travacalm following nineteen people being hospitalized and a further sixty-eight experiencing life-threatening adverse reactions as a consequence.[7] Quality issues famously delayed the launch of Clarinex (desloratadine) into the U.S. market when time to market was critical for Schering-Plough in advance of the patent expiry of Claritin (loratadine). The delay in launch due to manufacturing noncompliance at its Puerto Rico production facilities (and subsequent $500 million fine) meant that the company was denied an important window of opportunity to switch customers between the two brands—in effect, even before it was born, Clarinex was damaged by a quality issue.

There are various examples of brand strategy and communication errors in pharmaceuticals that have led to decline of the brand. One example is SmithKline Beckman's Tagamet and its failure to react to the aggressive marketing stance taken by Glaxo when it launched Zantac in the 1980s. Despite the profiles of the two product brands being very similar, Glaxo outthought and outresourced their brand to create the largest-selling medicine worldwide at the time. Another more recent example concerns Lipitor's launch by Warner-Lambert and Pfizer and its redefinition of the cholesterol market behind the efficacy strength of its product—despite many years of notice of its arrival, the incumbent companies and brands assumed that Lipitor's lack of long-term outcome data would protect their position—in reality, Warner Lambert and Pfizer redefined the market and sidestepped the issue until its own long-term data would become available.

Pricing within pharmaceuticals is a whole area of expertise in itself and is highly complex. There are examples in every company of brand decline being caused by either internal pricing decisions or external forces working on the pricing dynamic. In particular, failing to respond to generic pricing initiatives is a common brand strategy failure.

CONCLUSIONS

Many FMCG brands enjoy longevity that pharmaceutical marketers can only dream about. Maintaining a brand over time requires clear customer insight and management commitment to sustaining or nourishing a tangible (or intangible) benefit over time. The brand has to be led by a clear strategy, staying true to itself and its range.

There are numerous identified factors in the FMCG area that decide the dominance of brands versus distributor own brands (the equivalent to pharmaceutical generics). Pharmaceutical product brands generally possess a significant number of the advantages that would normally make long-term brand domination possible. The industry focuses huge attention on largely two—patent protection and safety. Pharmaceutical brands should enjoy a favorable position versus generics, but this is outweighed by the implications of patent expiry.

The causes that lead to FMCG brand decline, or brand death, are also well understood. Virtually all of them can equally lead to the decline of pharmaceutical brands, but again little enough thought is given to these peripheral subjects. Brand decline is largely driven by technology replacement, safety events, and brand strategy mistakes. Brand strategy mistakes can be numerous and varied, but include degradation in quality, communication errors, distribution mistakes, pricing problems, range confusion, the refusal to follow a durable market change, and, last, but not least, withdrawal of investment. Pharmaceutical brands suffer from virtually all the same factors of decline as FMCG brands.

Chapter 8

Sustaining a Product Brand
Over Time

In pharmaceuticals, the 1980s and 1990s saw easy and accelerating sales success in global pharmaceutical markets (see Table 8.1). At the time, there were enough significant products in the pipeline to sustain product brands post-patent expiry. The industry focused on blockbusters, with key measures such as the speed to the first $1 billion in annual sales being a crucial measure of success (as this increased the likelihood of achieving peak sales in the shortest possible time frame). This was compounded by the attitude at the time that nothing could be done about generics anyway, and the future was going to be led by research and development (R&D) productivity, sales and marketing critical mass, and aggressive patent defense.

The last years of the twentieth century, however, saw a change in attitudes, the genomics R&D boom hadn't arrived, new product development took longer and was more expensive than ever, and new breakthrough launches were being rapidly followed by close competitors (e.g., Vioxx and Celebrex, Actos and Avandia). But, individual companies were managing to sustain brand sales way past patent expiry—they were extending the life of their brands. The most extreme example being Wyeth's Premarin, which achieved peak sales fifty-nine years after its original launch in 1942.[1] Brands with staying power, therefore, do exist, as shown in Table 8.1. This has challenged some major preconceptions held in the industry.

The traditional view within the pharma industry is that the life cycle of a product brand is relatively short, and the death of the brand is inevitable due to patent expiry and the resultant rapid penetration of

Pharmaceuticals—Where's the Brand Logic?
Published by The Haworth Press, Inc., 2007. All rights reserved.
doi:10.1300/5836_08

TABLE 8.1. Product brands with staying power.

Brand	Company	Year of first launch	Year of peak sales	Longevity (time to peak)
Premarin	Wyeth	1942	2001	59
Augmentin	GSK	1981	2001	20
Sandimmun	Novartis	1980	2000	20
Humulin	Lilly	1982	2000	18
Toprol	AstraZeneca	1975	Not reached	27+
Depakote	Abbott	1983	2001	18

Source: Adapted from Scrip PJB publications "Can products with staying power be identified?" April 2004.

branded sales by generics. In the context of sustaining a brand over time, the concept of product life cycles (PLCs) in the pharmaceutical industry is worth considering. The PLC concept is a "none" industry-specific classic marketing tool, originally described in the 1950s (see Figure 8.1). This tool is much used and also much derided across varied industry sectors.

The best explanation of how the concept of PLCs might be relevant to pharmaceutical product brands was presented by Mick Kolassa, then associate professor of pharmaceutical marketing at the University of Mississippi. Mick's definition of a life cycle management plan is:

> A long-term brand plan designed and implemented to achieve the maximum brand potential within the context of the market life cycle and the life of the therapy/technology.[2]

The central premise was that discussions of PLCs were only relevant to markets and technologies and not to individual products themselves. A point that runs counter to what many pharmaceutical marketers are taught at the traditional early product managers courses they attend when starting out. Traditional logic suggests that individual pharmaceutical brands reach a peak in their sales approximately five years after launch and then start to decline. However, the facts and data don't seem to bare this out, with many products only hitting global peak sales at patent expiry in the U.S. (and EU) markets. According to the Spectrum report from Decision Resources (March

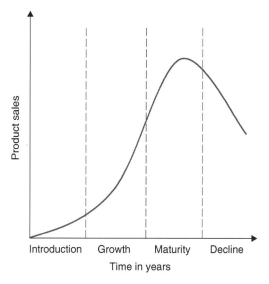

FIGURE 8.1. Product Life Cycle Diagram.

2004), reported in Scrip,[3] the longevity (time to peak sales) of pharmaceutical product brands has been declining recently and is now around twelve years. The report anticipates that this will remain relatively constant until approximately 2010, when, in fact, longevity will start to rise again due to the increasing number of biologics becoming available.

The PLC concept, when translated to pharmaceuticals, is often taught in a way that is too rigid and can lead to dubious conclusions. Some other common industry misconceptions that also exist include that prescription products live and die purely as a function of the patent, that once a product starts to decline it can never recover, that as your brand enters maturity, advertising and promotion (A&P) spending should be withdrawn. Also that line extensions should be reserved for when a product starts to age, and last, but not least, patent expiry of one brand in a class affects all the class.

In reality, sustaining a product brand over time, or extending the brand life, has a lot to do with good life cycle and brand management. Extending brand life requires a synthesis of many aspects of brand strategy and isn't just a matter of following the simple and basic principles inherent in the classic and erroneous product life cycle.

FIGURE 8.2. The pharmaceutical product life cycle. (*Source:* Adapted from Mick Kolassa presentation titled "Pharmaceutical life cycle management," August 2003.)

Although, aspects of the PLC do hold true for pharmaceutical life-cycle management when they are analyzed further, few brands are launched into the introductory phase; most are, in fact, launched into growth or mature markets where the battle is for current (and future) positioning, key opinion leader endorsement, market understanding, and the necessary sales resources to compete. Different tactics are appropriate in the different stages, and according to Kolassa, we should be tracking customers as much as sales to validate the PLC model (see Figure 8.2), as diffusion and adoption of the brand within the customer base is the key to implementing the correct tactics.

PHARMACEUTICAL BRAND LIFE DETERMINANTS

Procter & Gamble management does not believe in the product life cycle concept. They argue that in the consumer area a well-managed brand can last forever. The sales statistics of soap powder brands in a perennially mature market appear to bear this out. At the other extreme are the life cycles seen in the fashion industry, often only a season, never mind a whole year. Generally speaking, the consumer world appears to have a shortening of PLCs due to rapidly changing technologies. Innovations in chip technology have revolutionized phone, television, and computer design. The consequence of this is that product innovation becomes the key to future success, something very much in common with pharmaceuticals. Having said that, the key difference is the length of time required to innovate. A new mobile phone may have an R&D timeframe of just six months, whereas a new chemical entity (NCE) in pharma can take a decade and just a new formulation, two to four years. Patents do play a part in fast-moving consumer goods (FMCG) research and development but, generally speaking, their importance is less significant due to the product life cycle being shorter than litigation, where patent court bottlenecks in many countries are lengthy.

On the pharmaceutical side, basic patent protection lasts twenty years from the patent filing date in the United States, E.U., Japan, and other relevant territories where intellectual property (IP) is recognized and applied for. There is one exception to this rule: products filed before June 1995 in the United States have a patent duration of twenty years from filing or seventeen years from grant date, whichever gives the longer duration. In the majority of major markets, this can be further extended via supplementary product certification (SPC) or patent term extension (PTE) in the United States. The maximum extension is five years, and the calculation depends on the delay in obtaining the marketing approval and has to be assessed for each product.

The maximum patent protection is, therefore, likely to be twenty-five years from filing of the original patent and does not necessarily include potential marketing exclusivity of six years post approval, which is another term used in some markets like Japan. Common wisdom within the industry, however, is that the real time available to maximize sales is often as low as only seven years due to the R&D and registration process eating into the exclusivity period.

Looking at the same issue from the opposite direction and using the Decision Resources time to peak sales estimate of twelve years, other assumptions can shed light. If we assume this spells a definitive reduction in R&D spending and a phased reduction in A&P across global markets, the calculation leaves us with a viable estimate for a pharmaceutical product brand life cycle of perhaps fifteen years.

Posing the question "Is this different to FMCG brands?" can provide surprising conclusions. In fact, pharmaceutical product brands enjoy a longer effective brand life to benefit from branding activities than many brands within the high-tech and consumer world.

Branding of consumer high-tech products is a critical success factor for those operating in that dynamic area, and has led to major innovation in how to approach the issue. Intel is a perfect example, a brand name that has now become a household name through a simple advertising tactic of creating ads in conjunction with the companies that use its semiconductor chips. The "intel inside" logo and simple accompanying tune are known worldwide. In the early 1990s, no one was interested in what chips powered their PCs but only a decade later one of the most important buying criteria for a computer purchaser is the brand name as well as the speed of the chip inside. Doubters of the importance of branding within pharmaceuticals should realize that other industries, with shorter brand life cycles, have innovated and created real value through consistent and well-thought-through strategic branding. It could be argued that Intel is, in fact, a corporate range brand, so investment will be recovered over coming decades—but the same argument would apply to pharmaceuticals if the corporate brand name was used more extensively.

Analysis of the global pharmaceutical market and the brands within it shows clearly that each individual product brand will have its own brand life. Brand life is the length of time an individual product will remain both a significant sales contributor and an actively managed asset of the company in question.

One example shows some of the complexity. Pfizer's antiepileptic brand Neurontin (gabapentin) was originally launched in its first indication of epilepsy in the U.S. market in 1991. It later achieved a relatively rapid subsequent launch in the United Kingdom and Germany, and then suffered some delays before obtaining reimbursed launch in other parts of Europe such as France and Italy. It has subsequently been rolled out around the world but, believe it or not, is still not marketed in Japan in mid-2005. There is the distinct possibility that Neurontin will enjoy marketing exclusivity of six or more years in the world's second largest market after its U.S. patent expiry date is long gone.

This means that the Neurontin global brand manager, in 2003-2004, should in theory be supporting introduction in Japan, growth in all markets due to its strengthening data in the area of neuropathic pain, maturity in its core epilepsy indication where newer products such as Keppra are taking share, and at the same time preparing for decline in the United States and most EU countries due to imminent patent expiry! Added to this complexity is the fact that Neurontin has turned out to be a better pain product than antiepileptic over the years and will see a different rate of erosion of sales at expiry in different indications. Neurontin's chronic epilepsy sales will be significantly more resistant to generic penetration, because seizure free epileptics don't take risks with changing medication when lifestyle issues like ability to drive are at stake. Its intermittent and more acute usage as treatment for various types of neuropathic pain could well see more rapid substitution encouraged by Health Maintenance Organizations (HMOs) on the other hand. To further complicate matters, A&P will be under pressure as the follow-up compound Lyrica (pregabalin) was launched in the United States and EU during 2004 and the long-term future for the Pfizer AED portfolio needs to see a transfer from Neurontin to Lyrica, in particular in its pain indications.

This single example shows that the life of a pharmaceutical product brand is affected by numerous unique factors that make brand management extremely complicated. At the same time, it illustrates a critical challenge for the future of the pharmaceutical industry. The Neurontin example merely skimmed the surface of the level of complexity showing that launch rollout timing, usage in different indications, and the launch of follow-up compounds all impact on decision making. To these factors can be added, for example, the product status (e.g., biological versus prescription only versus OTC), that the different presentations of the same brand will erode at different rates (oral tablets versus powders versus solutions, versus injectables), and this still leaves the possibility of combining the original brand with another chemical entity to claim further patent protection as has been managed by other manufacturers. Advair/Seretide is a combination of Serevent and Flovent/Flixotide from GSK, and the new brand Vytorin is a combination of Zocor and Zetia in a deal between Merck and Schering-Plough.

It is, therefore, better to think of the length of a pharmaceutical product brand's life, rather than discuss the PLC, as it removes many of the constraints of a rigid model. The next section will look in more

TABLE 8.2. Brand life determinants.

Global launch timing
Global core market characteristics
Product type (biological, Rx, OTC, or combination)
Indications used in (acute, chronic)
Presentation usage (galenics)
Product brand naming strategies
Corporate strategy and commitment

depth at this aspect of what determines the length of a pharmaceutical brand's life.

The Neurontin example introduced some of the complexities of trying to assess how long a pharmaceutical product brand will thrive in a global market. Table 8.2 sets out brand life determinants—those factors that will have the biggest impact on the potential life of an individual medicine. These factors range from the timing of global launches, the different characteristics of the core markets, the type of pharmaceutical (as OTC brands have a different brand life to purely prescription [Rx] products), the different indications that a particular brand is used for as well as their different presentations and, finally, how the brand fits into corporate strategy and its resulting financial commitment to sustaining the brand.

Global Launch Timing and the Resultant Marketing Exclusivities

At the macro level, a clearly identifiable sequence of brand launches has become apparent over the last two decades in the major pharmaceutical markets (the major markets being the United States, Japan, Germany, United Kingdom, France, Spain, and Italy). This group of seven make up over 80 percent of the total branded global market irrespective of occasional category variations, explained by whether the big-selling brands have recently lost patent protection or not. At the end of 2004, measured in U.S. dollars, the top-seven markets accounted for $316 billion in IMS Health audited sales with the U.S. market alone being $173 billion and the second largest country being Japan with $57 billion.[4] They were followed by Germany, France, the

United Kingdom, Italy, and then Spain, respectively. Canada and Mexico come next, closely pursued by China.

The launch sequence is driven by a number of factors, including the need to target the biggest U.S. market first (it also has the most transparent and, therefore, fastest regulatory environment). The speed of subsequent reimbursement in Europe postregistration with the EMEA and the difficulties of trying to bridge clinical trial data between the West and Japan sometimes lead to extended lag periods around the world. As a result of this, the established norm for big product brand registrations and subsequent launches is:

1. U.S. launch
2. United Kingdom and Germany (usually within a year of U.S. launch)
3. Italy, Spain, and France (reimbursement delays can be from one to three years)
4. Japan (delays one to fifteen years)

The remaining major blocks that make up the world market are the rest of the European community (especially now accession has added another ten countries to the previous 2003 count of fifteen EU states), ASEAN countries (where regulatory convergence is making some belated progress), and then South American countries, in particular, Mexico and Brazil.

One of the key factors identified as critical to rapid sales growth of new product brands at the end of the 1990s was the percentage of the world market reached within two years of first launch. This is de facto a measure of success for the regulatory departments operating globally. McKinsey analysis conducted during 2000[5] suggested that some top pharma companies (MSD, Novartis, Pfizer, Glaxo Wellcome, and Lilly) could access just under 70 percent of the world market in that two-year timeframe, while other big players like BMS, SB, and J&J achieved only 50 percent access. The average number is likely to shift upward in the future, as the regulatory environment continues to consolidate through consultation and agreement of standards. But this speed to market is difficult to replicate in smaller organizations and will remain a significant factor in the future. Rapid global registration takes a huge amount of resources to achieve.

Figure 8.3 shows some previous analysis of IMS Health data from MAT Q3 2000—it shows that there were considerable differences between the brands making up the top twenty global brands and the top twenty in the United States, EU, rest of the world (ROW), and Japan.

The U.S. list is pretty well representative of the global top twenty at the time. Something you would expect due to the size of the U.S. market and speed of registration of new products. The EU and the rest of the world lists do have similarities in that the progression of NCEs can be seen, and a pattern is recognized. But, by the time we look at Japan, the differences are significant. Within the last four years, there has probably been a narrowing of the gap between the two extreme markets, the United States and Japan, but big differences will remain for some time to come, particularly in those disease areas where clinical study end points are not easily measured. A phenomenon that makes bridging studies, for instance, difficult to predict in areas such as CNS. Until March 2005, no CNS bridging study has been accepted, despite numerous attempts by the top companies.

Even within the EU, a more detailed market-by-market analysis shows significant differences (see Figure 8.4). The top twenty brands

Global rank brand	U.S. rank brand	EU rank brand	ROW rank brand	Japan rank brand
1. Losec	1. Losec	1. Losec	1. Lipitor	1. Mevalotin
2. Lipitor	2. Lipitor	2. Lipitor	2. Losec	2. Gaster
3. Zocor	3. Ogastro	3. Zocor	3. Voltaren	3. Norvasc
4. Norvasc	4. Prozac	4. Norvasc	4. Zocor	4. Zocor
5. Prozac	5. Zocor	5. Pulmicort	5. Norvasc	5. Epogin
6. Ogastro	6. Epogen	6. Seroxat	6. Renitec	6. Cravit
7. Seroxat	7. Celebrex	7. Flixotide	7. Celebrex	7. Uft
8. Zyprexa	8. Zyprexa	8. Sandimmun	8. Seroxat	8. Panaldine
9. Zoloft	9. Claritine	9. Augmentin	9. Augmentin	9. Enantone
10. Erypo	10. Zoloft	10. Serevent	10. Adalat	10. Harnal
11. Claritine	11. Seroxat	11. Ciproxin	11. Zyprexa	11. Basen
12. Epogen	12. Erypo	12. Zyprexa	12. Viagra	12. Omnipaque
13. Celebrex	13. Glucophage	13. Renitec	13. Pravachol	13. Espo
14. Augmentin	14. Norvasc	14. Pravachol	14. Ventolin	14. Sector
15. Risperdal	15. Augmentin	15. Erypo	15. Ciproxin	15. Adalat
16. Ciproxin	16. Risperdal	16. Prozac	16. Klacid	16. Amlodin
17. Glucophage	17. Zithromax	17. Voltaren	17. Xenical	17. Iopamiron
18. Renitec	18. Vioxx	18. Cipramil	18. Zoloft	18. Renitec
19. Pravachol	19. Ciproxin	19. Zantac	19. Flixotide	19. Selbex
20. Zithromax	20. Neurontin	20. Haemo Glu	20. Zantac	20. Cefzon

FIGURE 8.3. A global pharmaceutical launch sequence; leading pharmaceutical brands by major market (2000). (*Source:* Giles D. Moss analysis of IMSHealth retail sales and hospital sales where available. MAT Q3 2000.)

Top 20 rank	Top 20 European Union (combined)	United Kingdom (Retail & Hospital)	Germany (Retail & Hospital)	France (Retail)	Italy (Retail & Hospital)	Spain (Retail)
9	Augmentin	Adalat	Actraphane	Augmentin	Adalat	Almax
18	Cipramil	Becotide	Betaloc	Di Antalvic	Ativan	Aspirin Bay
11	Ciproxin	Cardura	Ciproxin	Doliprane	Augmentin	Augmentin
15	Erypo	Effexor	Duragesic	Efferalgan	Aulin	Decapeptyl
7	Flixotide	Flixotide	Enantone	Flixotide	Cardura	Disgren
20	Haemo Glu	Lanzo	Glucometer	Lipanthyl	Ciproxin	Lipitor
2	Lipitor	Lipitor	H.B. Vax	Lipitor	Enantone	Losec
1	Losec	Losec	Haemo GIU	Losec	Erypo	Norvasc
4	Norvasc	Norvasc	Insumank	Norvasc	Fortum	Plavix
14	Pravachol	Pravachol	Lipitor	Plavix	Lipitor	Prozac
16	Prozac	Prozac	Lipobay	Pravachol	Losec	Pulmicort
5	Pulmicort	Pulmicort	Losec	Prozac	Macladin	Risperdal
13	Renitec	Ranitidine	Neupogen	Pulmicort	Norvasc	Sandimmun
8	Sandimmun	Serevent	Norvasc	Pyostacine	Ranidil	Serevent
10	Serevent	Seroxat	Plavix	Serevent	Renitec	Seroxat
6	Seroxat	Ventolin	Pulmicort	Seroxat	Rocephin	Vioxx
17	Voltaren	Zestril	Sandimmun	Tanakan	Sandimmun	Voltaren
19	Zantac	Zocor	Taxol BMS	Vastarel	Vaseretic	Zantac
3	Zocor	Zoladex	Twinrix	Vasten	Zantac	Zocor
12	Zyprexa	Zyprexa	Zocor	Zocor	Zocor	Zyprexa

FIGURE 8.4. An European pharmaceutical launch sequence; analysis of top 20 EU brands (2000). (*Source:* Giles D. Moss analysis of IMS Health retail sales and hospital sales where available. MAT Q3 2000.)

were listed in alphabetical order to try and establish a recognizable pattern below. Augmentin, for instance, although the ninth selling brand within the overall EU, was not a top-ranked brand in either the United Kingdom or Germany. Similarly, Flixotide, ranked seventh in the fifteen EU member states, was not yet successful in France, Italy, and Spain at the time of the analysis. With the enlargement of the EU to twenty-five countries in May 2004, there will be even more variation in launch timings despite the concept of a common regulatory approval date.

The fact that a launch sequence exists, and will continue to be the case for decades to come, impacts on the strategic direction a company can take when building their product brands and, indeed, their brand portfolios. The pharmaceutical world tends to focus almost exclusively on the U.S. market, due to its weight in world market terms. But, because of this, significant brand strategy mistakes can be made by senior management obsessed by the scale and rapid returns of the American market. As a result, it is common for global product management to struggle to optimally support launches at different times in different markets, something that in turn impacts on product brand life.

Global Market Characteristics

The sequence of launch timing has already shown major inter-market variation, but other factors also impact on the sustainability of a product brand over time. In the broadest terms, the U.S. market tends to see rapid access and launch, but also rapid decline when brands are opened up to generic competition. The United Kingdom (U.K.) and Germany behave in a fashion largely similar to the United States, except that the U.K. physician population is famous for slow adoption of innovative products and, therefore, launch success curves are always disappointing in comparison to other markets. Italy, Spain, and France tend to see slower access to market, generally with France being the slowest unless the product brand in question is marketed by a French company. On the flip side, these three more Latin countries also have tended to see slower generic penetration when a brand loses its patent protection. Significant changes are occurring, however, such as the cuts in both price and reimbursement status in Italy, and, the at times unusual public statements and behavior of numerous governments. The French measures include reaching agreement with the physicians, in June 2002, to pay more per consultation in return for a change in prescribing generics more regularly, and even being prepared to write to each patient showing a breakdown of their con-sumption and exhorting them to choose generics whenever possible.

Convergence in government policy and activity, with the aim of in-creasing the wider usage of generics, will reduce some of these mar-ket differences over time. Despite the fact that all governments are under extreme pressure to reduce health care spending, each country has a culturally different approach to how it delivers its health care, and, as such, important opportunities for sustaining a brand will re-main exploitable in the future.

Japan is where the differences are really accentuated; registration delays can run into decades rather than years, with companies often having to rely on marketing exclusivity rather than patent protection at launch. On the plus side, at present, generics don't really play a part in the Japanese market and, therefore, long-term sustainability of a well-marketed brand can be lengthy.

The rest of Asia Pacific offers incredible variety in the market characteristics, ranging from prescription markets with full reimbursement but low prices (Australia and New Zealand), to prescription markets with free pricing but where the patient pays everything (Indonesia and Philippines), to countries where the physician dispenses and where copayment exists for those lucky enough to have good employment benefits (Malaysia and Singapore). A number of these other countries were characterized in the past by IP rights being flouted (e.g., China and India), and, as a result, the key becomes competing in the market versus local generic manufacturers, as well as maximizing sales while some protection still exists. Long-term-sustainable sales can then be achieved, but probably at a cost of slower initial sales uptake and an overall smaller return due to the competitiveness of the market.

Over the coming decades, regulatory convergence will even impact on Japan, bringing it closer to the regulatory timeframes seen in the West. Therefore, the current differences caused by global launch timing sequences and different market characteristics will, to some extent, diminish. But, despite this, there will still be many opportunities to sustain your brand. To achieve this, a company will have to be prepared to adopt strategic branding rather than the current tactical approach seen so far and change the mindset that is dominated by U.S. market timelines.

Product Type

There are a multitude of different types of product brands that tend to be lumped together when discussion on pharmaceutical branding takes place. The major categories of product brands include traditional prescription-only products, whether they are for acute or chronic diseases, over-the-counter (OTC) medicines, which can be freely purchased with limited restrictions, and biologicals, which are again a prescription-only product but are distinct in their makeup and their method of production. The final major category is represented by new brands that are a combination of two previously marketed chemical entities allowing, in most instances, additional regulatory exclusivity.

According to PharmExec, who analyzed 2002 global sales, the top-ten global brands (see Figure 8.5) showed three biologicals within the list. In compiling the list, they classified together a number of individual brands for Erythropoietin, Interferons, and human insulins from numerous markets around the world to come up with this top-tier. The number of biologicals will be expected to increase in the future, especially if moves to register biogenerics prove to be slow; they will be around longer than the other prescription brands without competition. The figure looks at the top-ten 2002 product brands and then categorizes them according to chronic (long-term) or acute (short-term) usage and the level of prescribing influence of primary care physicians (PCPs) and specialists. To this top-ten list were added Crestor, Effexor, and Paxil/Seroxat as candidates at the time of the analysis, likely to figure in the top ten in the future.

Prescription brands can present quite different opportunities for sustaining the brand over the long term, dependant on whether they are used to treat chronic or acute diseases and whether they are pri-

	Global GP (chronic)	Global GP (acute)	Global specialist (chronic)	Global specialist (acute)
Lipitor	+++		++	
Erythropoietin		+		+++
Zocor	+++		++	
Interferons		+		+++
Prilosec/Losec	+++		++	
Human	++		++	
Zyprexa	++		+++	
Norvasc	+++		++	
Prevacid	+++			++
Celebrex	+++		++	++
Crestor	+++		++	
Effexor	+++	+++	++	++
Paxil	+++	+++	++	++

Majority of prescribing: +++   Exclusive prescribing influence: +++
Repeat prescribing: ++   Influential prescribing influence: ++
Minor prescribing: +   Little prescribing influence: +

FIGURE 8.5. Blockbuster classification.

marily prescribed by PCPs or specialists (hospital or privately based specialists/consultants). In the case where a particular disease or symptom is treated for a relatively short period of time (an acute illness such as an infection), the opportunity to switch both the patients' and physicians' loyalty from a branded product is relatively high. There is the general perception that short-term versus life-long medications are quite different. Chronic treatments are those life-long treatments for conditions such as high cholesterol or blood pressure; they tend to increase the chances a product brand will be used long after a generic copy becomes available. The other significant factor is that the specialist community has tended to maintain its independence of prescribing rights during a time when PCPs have seen erosion in their ability to prescribe as they wish. HMOs and government payers around the world are insisting on tighter control of prescription items, both for the indications they are used for, and in what quantity. The figure gives some clues about which types of product are most resistant to generic substitution, irrespective of the local market conditions. One easy hypothesis would be that those brands whose prescription is largely controlled by specialists will enjoy a greater chance of extending their brand life, particularly those whose manufacture makes them difficult to copy (such as biologicals). Those product brands, with these advantages, will be easier to sustain over time and would, therefore, benefit more from strategic brand management at an early stage in their brand life.

Prozac (fluoxetine) represents a good case study of a product brand that became a cultural icon during the 1990s following its launch in 1987. The fundamentals are that, in depression, six months of treatment is recommended, but, in reality, the average length of a script for patients (looked at in longitudinal databases) is a lot less, at approximately three months. Over time, the main prescribers of Prozac, and the other SSRIs, became PCPs rather than specialists. This meant that the vast majority of Prozac sales were coming from a disease area with a short (acute) length of treatment and from prescribers dominated by their HMO or governmental recommendations. The Prozac brand was and still is extremely strong and, in this context, it may have been thought possible to reduce the impact of generic penetration due to patient brand loyalty.

But when Prozac did come off patent in the United States, it saw very aggressive penetration of generics losing up to 80 percent of its sales within a matter of weeks.[6] This may be explained to some extent by the notoriety of the brand and the way the popular press around the world adopted it. While its manufacturer, Lilly, was trying to deal with issues management (positive sales growth in some countries due to consumer demand and negative reaction from physicians in other countries where there was a backlash against such a popular consumer brand), the press and managed care made it their responsibility to ensure wide awareness of cheaper fluoxetine becoming available. A great deal of effort was also made by HMOs to ensure that patients were briefed in advance about the availability of cheaper fluoxetine, allowing a rapid switch at launch. Another direct impact on Lilly, due perhaps to product management and senior management distraction, was that Prozac's clinical development program was late. Although Lilly managed to register depression initially, followed by bulimia, geriatric depression, and finally premenstrual disorder before U.S. expiry, they failed to get value from panic disorder, post-traumatic stress disorder (PTSD), and childhood depression while still protected in the world's biggest market. Prozac prescribing was, therefore, quite concentrated in the depression indication, and the other indications had not been promoted long enough to give it a long-term sustainable brand presence. The huge media interest made Prozac an icon and then destroyed the branded sales at expiry.

In sharp contrast, prescription brands that treat long-term (chronic) illnesses can benefit from brand loyalty years after losing their patent protection. This is especially the case in serious illnesses such as uncontrolled hypertension, angina, or heart failure, where small variations in bioavailability of the product brand in question can create a significant change for the previously stable patient. Norvasc (amlodipine) is a calcium channel blocker (CCB) that is often used in unstable heart disease cases and, therefore, to some extent should be expected to benefit from reluctance to switch to generic when its patent expires. If Norvasc had additionally been a specialist-only brand, then its chances of sustainability would be increased. As an analogue for Norvasc, another older CCB called Tildiem (diltiazem), a brand for angina, has been recommended to be prescribed by brand in the United Kingdom, despite the presence of numerous generic alternatives.

This necessity to maintain control is also easily understood in diseases that significantly impact on quality of life, such as epilepsy. Once a patient is a diagnosed epileptic, their driving license is removed in the vast majority of countries around the world, and this has a major impact on their independence and quality of life. An epileptic patient needs to have been seizure free for a significant amount of time before they can get their license back (six months or more commonly one year). As a result, any change in medication becomes a risk for a seizure-free person, which many will be unprepared to take. The aged epilepsy products carbamazepine and valproic acid are excellent examples of this phenomenon; carbamazepine still holds onto more than 60 percent of its original brand share as Tegretol (Novartis) in the United States. Valproic acid (Epilim from Sanofi-Aventis) did even better in the United Kingdom with greater than 80 percent of sales retained as branded despite having been off patent for more than thirty years.

Over-the-Counter (OTC) Brand

OTC brands are pharmaceutical product brands that can be purchased in a pharmacy or other retail environment such as a supermarket (dependant on individual country legislation). The patient may be able to just pick the brand off the relevant shelf or might need to request an item from an area supervised by a pharmacist. The advantage of taking a prescription brand OTC is that the brand can live for ever, as can any consumer purchasable brand, and therefore a longer-term view can be taken.

The most frequently used pharmaceutical and OTC products around the world in 2000 are shown in the Tables 8.3, 8.4, and 8.5 and, as expected, are dominated by OTC brands. Again, significant variations are seen between the United States, E.U., and Japan, perhaps more greatly influenced by local culture and local regulatory decisions than the more regulated prescription-only brands. It illustrates quite simply that brand sustainability can be influenced in different ways in different markets and at different stages of a brand's life.

In the United States, after Ventolin and Voltaren (excluding mouthwashes, hydrogen peroxide, and ethanol), aspirin- and paracetamol-based brands make up a significant number of standard counting

TABLE 8.3. United States: Pharmaceutical and OTC brands by counting units.

	Product	Corporation	Molecule
1	Listerine	Pfizer	Benzoic Acid
2	Ethanol L.U.	Lab Unknown	Ethanol
3	Hydrogen Per L.U.	Lab Unknown	Hydrogen Peroxide
4	Salbutamol S-PI	Schering-Plough	Salbutamol
5	Mouth Wash L.U.	Lab Unknown	Benzoic Acid
6	Tylenol	Johnson & Johnson	Paracetamol
7	Acetyl S. Acid L.U.	Lab Unknown	Acetylsalicylic Acid
8	Isopropyl Alc. L.U.	Lab Unknown	2-Propanol
9	Advil	American Home	Ibuprofen
10	Glucophage	Bristol-Myers Squibb	Metformin
11	Renu	Bausch & Lomb	Boric Acid
12	Premarin	American Home	Estrogenic
13	Ibuprofen L.U.	Lab Unknown	Ibuprofen
14	Tums	SmithKline Beecham	Calcium
15	Synthroid	Basf	Levothyroxine Sod.
16	Azmacort	Aventis	Triamcinolone A.
17	Sod. Chloride L.U.	Lab Unknown	Sodium
18	Paracetamol L.U.	Lab Unknown	Paracetamol
19	Eye Drops L.U.	Lab Unknown	Naphazoline
20	Flixonase	Glaxo Wellcome	Fluticasone

units. Some premium brands exist within the market such as Aspirin from Bayer and Advil and Tylenol in the United States, but much of the volume sold is more like a commodity than a brand. Generally speaking, the OTC market is less attractive than the prescription market as, despite high volume, growth and prices tend to be lower. Promotion is more expensive due to the need to use mass media such as TV advertising, and a much bigger margin is taken by the pharmacist. Some notable failures have occurred, including the switch of the once global blockbusters Zantac and Tagamet. Surprisingly, these once revolutionary product brands were unable to differentiate themselves in the market as effectively as hoped for and failed to make a major success. This failure perhaps was due to lack of experience at the time of how to switch brands or perhaps because "claims" in the

TABLE 8.4. Europe: Pharmaceutical and OTC brands by counting units.

	Product	Corporation	Molecule
1	Ventolin	Glaxo Wellcome	Salbutamol
2	Voltaren	Novartis	Diclofenac
3	Vigantol	Merck Kgaa	Colecalciferol
4	Locabiotal	Servier	Fusafungine
5	Becotide	Glaxo Wellcome	Beclometasone
6	Berodual	Boehringer Ingelheim	Fenoterol
7	Aspirin Bayer	Bayer	Acetylsalicylic Acid
8	Pastiglie Valda	SmithKline Beecham	Cicliomenol
9	Efferalgan	Bristol-Myers Squibb	Paracetamol
10	Pulmicort	AstraZeneca	Budesonide
11	Betadine	Degussa-Huels	Povidone-Iodine
12	Atrovent	Boehringer-Ingelheim	Ipratropium Bromide
13	Inolaxol	Fournier	Sterculia Gum
14	Ativan	American Home	Lorazepam
15	Bricanyl	AstraZeneca	Terbutaline
16	Aqueous L.U.	Lab Unknown	2-Phenoxyethanol
17	Gaviscon	Reckit Benckiser	Alginic Acid
18	Salbutamol L.U.	Lab Unknown	Salbutamol
19	Amuchina	Amuchina	Na Hypochlorous A.

OTC area have not been rigorously policed in the past. Over claims from less efficacious products already occupy the ideal positioning in the mind of the consumer in many therapeutic categories and even if a new switch is clearly superior from a clinical point of view, the message can be near impossible to get across.

Many of the major "big pharma" companies run separate consumer health care divisions as has previously been discussed in Chapter 4, on corporate branding. What makes the OTC segment of the global business interesting is that many of the brand name companies are fighting each other through their consumer health care divisions or generic divisions in what appears to be an uncoordinated fashion. In theory, it is within big pharma's interests to create and maintain successful long-term product brands that can help to alleviate the currently frantic search for NCEs. Why then would Sandoz, generics

TABLE 8.5. Japan: Pharmaceutical and OTC brands by counting units.

	Product	Corporation	Molecule
1	Kary	Santen Seiyaku	Pirenoxine
2	Catalin	Senju Seiyaku	Pirenoxine
3	Tarivid	Santen Seiyaku	Ofloxacin
4	Isodine	Meiji Seika	Povidone-Iodine
5	Rohto	Rohto Corp.	Chlorphenamine
6	Sancoba	Santen Seiyaku	Cyanocobalamine
7	Hyalein	Santen Seiyaku	Hyaluronic Acid
8	Selbex	Eisai	Teprenone
9	Smile Lion	Lion	Potassium
10	Maskin	Maruishi	Chlorhexidine
11	Niflan Yoshitomi	Senju Seiyaku	Pranoprofen
12	Epadel Mochida	Mochida	Eicosapentaenoic A.
13	Flumetholon	Santen Seiyaku	Fluoromethonone
14	Osvan	Takeda	Benzalkonium Cl.
15	Germitol Maruishi	Maruishi	Benzalkonium Cl.
16	Biofermin	Biofermin Seiyaku	Streptococcus F.
17	Hibitane	AstraZeneca	Chlorhexidine
18	Cabagin	Kowa Shinyaku	Aluminium
19	Sante	Santen Seiyaku	Chlorphenamine
20	Gester	Yamanouchi Seiyaku	Famotidine

arm of Novartis AG, actively lobby the FDA for an approval of a generic version of an aging human growth hormone, which, due to its biological status, currently cannot be attacked by generics? Perhaps a clearer strategy would be to buy into a biological future and drive long-term shareholder value that way rather than encourage the move to biogenerics.

As far as governments are concerned, any cost they can shift to the consumer in a government-funded health care system is a positive, and, therefore, much debate continues concerning whether companies can be forced to move their product brands OTC. One recent example shows that the stakes can be high when it comes to moving a successful prescription product brand to OTC.

In the United States, the case study tends to be Claritin (loratadine) from Schering-Plough, an antihistamine brand that had extensively used direct-to-consumer advertising (DTCa) to create a strong brand image with American consumers. According to reports in 2000, Claritin had successfully transformed the allergy market, previously dominated by OTCs, into one in which 53 percent of allergy patients began to buy prescription products. This change had taken at least $248 million in eighteen months of DTCa activity, according to McKinsey.[7]

Unfortunately, Schering-Plough found itself in the position that Claritin might be forcibly taken OTC by the FDA without its consent (this didn't happen and to date the FDA has never forced a mandatory switch, unlike the European regulators). Initially opposed to the proposal, Schering-Plough later agreed in a surprise move to take the product OTC at the same time it was introducing its follow-up brand Clarinex (desloratadine). In theory, everything was prepared for a major prescription product brand with sales in the billions to be converted to OTC in a classic brand strategy move. In reality, the timing was poor, the market did not move quickly to OTC due to numerous factors, Clarinex was delayed, and Schering-Plough was plunged into a commercial and financial crisis that saw a total change of the top management team within a year.

The OTC situation is being further complicated by aggressive activity from some of the biggest players. In the U.K. market, GSK consumer health care took off-patent omeprazole (prescription brand Losec/Prilosec from AstraZeneca) through the switch process, something that, despite a deal with Procter & Gamble in the U.S. market to do exactly that, AstraZeneca had declined to do in the United Kingdom. This may have something to do with the fact that in the United Kingdom, the marketing exclusivity, which acts as a reward for taking the brand through the filing process, only gives three months protection before generics can take advantage too. In the United States, in contrast, the J&J MSD Consumer Pharmaceuticals joint venture will benefit from three years of marketing exclusivity when it finally switches another lipid-lowering product brand, Lovastatin 20 mg, after having had an earlier application for the 10 mg rejected. The same joint venture was also the first company to switch a statin in the United

Kingdom when it launched 10 mg simvastatin as the brand Zocor Heart-Pro during 2004.[8]

In addition to a more aggressive stance being taken over positive switching to OTC, there is pressure from independent organizations such as the Tufts Center for the Study of Drug Development. They predict in their "Outlook 2004" report[9] that the FDA will speed the switching of prescription medicines to OTC especially as fifteen blockbusters are due to lose patent protection in the U.S. market before the end of 2008. As a result, strategic brand management across the interface between different but co-owned pharmaceutical and consumer health care companies will be more critical in the future. The benefits could be enormous and long lasting for pharmaceutical product brands, but the way forward is still not clear.

Biologicals

The brand name biotech industry has been around for only just over twenty-five years, and so far has not had to contend with cheap generic copies of its major products. At present, this is due to the fact that biologicals are so new that they are still patent protected, but this will change in the next few years. A number of old brands such as Lilly's bioengineered insulin Humulin and Genentech's Nutropin growth hormone will lose their original patent protection early in the new century and may not be able to protect themselves from copies due to their relatively simple structure. The crux of the issue is the size of the molecules in question. Traditional chemical drugs are relatively small molecules and are produced through well-understood chemical processes, while biologicals are large, complex, and heavily dependent on the cultures that create them as far as impurities are concerned. Biotech-manufactured molecules are generally much bigger and more complex as they tend to be produced by splicing the genetic sequence of a targeted protein into living culture cells, which then create copies of the protein for later extraction. Due to the fact that different cell cultures in different lines can create proteins with potentially quite different side effects for the patient, regulatory authorities are finding it extremely difficult to decide if they can approve copies.

There are, of course, some nonbiological exceptions like Augmentin (clavulanate and amoxicillin), another combination from GSK. Augmentin fought off generic competition for years post patent expiry, with a combination of aggressive legal defense following theft of its proprietary strain, allied to the fact that there are not that many antibiotic production facilities around the world that could easily replicate the many steps in its synthesis.

Generics of traditional versions of drugs have a relatively simple procedure to obtain registration by being allowed to rely on the data generated by the original manufacturer and as such are considered to be virtually identical. Specific procedures were created in the different global markets to allow this to happen (e.g., U.S. congressional approval in 1984 for small molecules)—the problem is that these procedures do not currently cover newer biologicals, and the authorities are struggling with the issue of not being able to measure the degrees of similarity between original biologicals and their potential copies.

For the short term, biologicals are effectively protected from traditional pharmaceutical competition until science moves forward. This will last until advances are made as far as measurement is concerned or regulatory agencies are prepared to take significantly more risk in approvals than they are currently.

Biotech companies are aware that eventually the situation will change, with the simpler early product brands already switching patients to new formulations copying classic brand strategy. They hope to ensure that continued full-price sales are protected as regulatory changes put them at risk (e.g., Genentech's Nutropin).

The most successful biological, erythropoietin (Amgen), has also interestingly been marketed by different companies in different indications with different brand names around the world. Ostensively, the reason for this may have been the nature of commercialization for biotech companies, where they were forced by commercial pressures to out-license some separate target physician and patient groups (while retaining others for their own marketing operations). The result, however, is superb segmentation, for example, Procrit (J&J) and Eprex (Roche) in the United States, Epogen and Espo from Chugai in the European community, and Epogin (Kirin), Arnesp (Chirin), and Neo-Recormon (Sankyo) in Japan. Therefore, whether by chance or by design, another barrier to easy generic replacement has been achieved by the web of contracts, relationships, data protection, and use of different product brand names to allow commercialization around the world.

Biologicals, therefore, represent an excellent example of how pharmaceutical product brands can be sustained over a considerable amount of time and would benefit from a strategic approach to brand management.

New Brands That Are Merely Two Other Brands Combined

This phenomenon is perhaps becoming more common as companies realize that with fewer NCEs reaching the market, there is an added incentive to get the most from their current portfolio. Recent examples include the combination of Pfizer's cardiovascular brands Lipitor and Norvasc/Istin. The new combination brand, named Caduet, combines the world's top-selling statin and the top-selling calcium channel blocker (in 2003) in a single pill containing the active chemical entities atorvastatin and amlodipine. Similarly, two recent GSK examples continue its long tradition of combined product brands. Its triptan Imitrex (sumatriptan) is being developed in phase 3 as a combination with naproxen, to be called Trexima when it comes to the market. Secondly, GSK has Andamet (rosiglitazone and metformin) to complement Avandia (rosiglitazone), combining with an off-patent BMS brand originally called Glucophage, which has been available to treat the type 2 diabetic patient for some time.

Various earlier analogues, of previously created combination brands, exist in the cardiovascular area, stretching over a number of decades. Although many met with inconsistent success, they created a firm product brand tactic for improved sustainability, e.g., Tenormin (atenolol) being combined with nifedipine (brand name Adalat) to create Tenif, and, similarly, Tenormin being combined with a generic thiazide diuretic in the branded form of Tenoretic. This model was then followed by a host of angiotensin converting enzyme (ACE) inhibitor product brands in the 1990s, which were similarly combined with thiazide diuretics in a single pill—Capoten (generic captopril) and the combination Capozide (captopril and hydrochlorothiazide); similarly, Zestril (lisinopril) and Zestoretic.

The idea, therefore, isn't new, but recent court action has started to bring into question the basis of whether patent protection will remain in the future. The success of GSK's asthma franchise has been built on the regular combination of previously marketed brands as they approach patent expiry. Massive marketing budgets are then made available to switch asthma sufferers between the two product brands to maintain franchise success across decades. A British High Court decision in March 2004 brought into question the validity of the patents protecting GSK's Seretide/Advair brand containing salmeterol and fluticasone propionate. A number of generic companies had challenged the key patents for Seretide on the grounds of "obviousness." They argued that the combination of Flovent/Flixotide (fluticasone propionate) and Serevent (salmeterol) was not a truly inventive step and was, in fact, entirely "obvious." Although the United Kingdom represents a relatively small percent of global sales for the $4 billion seller in 2003, generic companies such as Cipla and Ivax could well raise the risk of patent challenges in more markets around the world and change the regulatory environment for this particular tactic.

A new brand that combines a completely new chemical entity with an older established brand would, however, be a lot more difficult to attack. This tactic has recently been employed by MSD in their combination of Zocor (simvastatin) and Zetia (ezetimibe) from Schering-Plough. The rationale behind the treatment combination is increased efficacy, to allow the new brand Vytorin to compete effectively against Lipitor (atorvastatin). Combining two chemical entities, which work in different but potentially synergistic ways, as well as improving the only real weakness of Zocor, offered a superb antigeneric tactic for the second biggest selling brand in the world at the end of 2003. The new brand will be promoted in the United States by a Merck/Schering joint venture and elsewhere largely separately.

As can be seen, there are numerous opportunities for combining an existing successful brand with another chemical entity, preferably a patent protected one, as it ages.

Indications a Product Brand Is Used

A single pharmaceutical product brand may have more than one disease state in which it is found to have utility. Sometimes, the diseases in question are relatively closely related. ACE inhibitors tend to enjoy indications in both hypertension, prevention of myocardial infarction (MI), and disease progression in left ventricular dysfunction, as well as cardiac failure due to the involvement of acetylcholines-terase in various cardiac mechanisms. Losec (omeprazole), for instance, has numerous indications that are relatively closely related such as esophageal reflux disease, duodenal and benign gastric ulcers (including those complicating nonsteroidal anti inflammatory drug (NSAID) therapy), and eradication of *Helicobacter pylori* (in association with antibiotics) in peptic ulcer disease. A third example already covered in this chapter is Prozac, with numerous different indications achieved during its brand life.

On other occasions, greater distinction in disease usage is observed, particularly where the disease in question stems from the central nervous system (CNS), where there are many overlapping pathways, and possible side effects of drug activity within the brain.

The impact of these different uses can have an effect on the sustainability of the brand in question. Neurontin (gabapentin) is active within the brain and has already been used as a simple example where its use in quite different indications will impact the speed of decline to generics in those different disease states. Many aspects of research and development are based on serendipity, and, in the case of Neurontin, the U.S. affiliate of Warner Lambert (Pfizer later bought the company to acquire Lipitor) had enough money in the budget for one major study in an alternative indication to epilepsy. The marketing team was asked for its best guess to maximize the potential future sales, and they plumped for neuropathic pain. To some extent, the rest is history; the open-label study showed efficacy, and this along with additional studies started to drive incredible off-label sales growth in the U.S. market. In the end, an official indication for pain was given in 2002, when at the time greater than 80 percent of U.S. sales were written off label. It could be argued that indication approval was more to save face for the FDA than because of the quality of the pivotal clinical trials presented in the file, but that doesn't matter. The result was use of Neurontin in a wide variety of both epilepsy and pain states that will greatly benefit its sustainability as patent expiry starts to bite.

Lyrica (pregabalin) is the Pfizer follow-up brand for Neurontin, and Pfizer learnt its development lessons here from the Warner Lambert days immediately. The varied CNS activity of the chemical entity was recognized early on, in part due to the need to replace Neurontin sales in a myriad of off-label uses. Development was initiated in a number of indications early in the development process.

This tactic is a classic brand strategy for the new century. The size of Pfizer at greater than $40 billion annual sales means that if a brand is going to be worth focus from the corporation's sales and marketing resources, then its peak sales potential has to be at least a billion. The traditional route to market for AEDs is through the epilepsy indication, a disease with relatively easily defined end points. Registration, therefore, as an add-on epilepsy agent provides market access but provides a relatively small sales potential due to the global sales potential in epilepsy only being approximately $1.8 billion at the end of 2003. Rather than conducting open-label studies with interested investigators, as was the case with Neurontin, Pfizer took the gamble of separate simultaneous development plans in epilepsy, neuropathic pain and generalized anxiety disorder (GAD).

The multiple launches under a similar brand name (presumably a request for multiple brand names was rejected) pose an unprecedented marketing brand challenge—to differentiate the brand at the same time in three separate disease states, with overlapping but characteristically quite different target physicians, at the same time without causing confusion internally or externally.

Presentation Usage

Development of different galenical forms, or presentations, is another very effective way of sustaining a product brand over time. This brand tactic is effective at any time in the brand life but tends to be reserved for when the brand is starting to age. Often, the effectiveness of any particular tactic is heavily dependant on the intellectual property status of the brand in question and the technology being used for the new presentation.

When looking at different presentations, it is also worth noting there is a vast array of ways of delivering an active ingredient, and, according to Vision Gain and their 2004 report "The future of drug delivery,"[10] the global market for drug delivery technologies reached $28 billion in 1998 and is forecast to grow to $78 billion by 2005. Examples include polymer-matrix-controlled release systems, liposomal delivery, osmotic pumps, nasal sprays, skin patches, and parenteral control release systems, among others.

During the 1980s and 1990s, a number of companies created profitable business models by securing patents on these novel delivery systems. In return for a royalty, the delivery system company would, for example, encapsulate, deliver via an inhaler, or reformulate an active chemical entity that was already on the market. The new presentation would have protection from generic "like-for-like" competition when the original product brands patent expired, and heavy sales and marketing investment was given to allow conversion of prescribers and patients to a new presentation in advance of patent expiry. This tactic effectively increases the patent life of that presentation of the original brand. The IP landscape is now changing with most of the original patents on these delivery systems expiring in the first five years of the new millennium. In the mean time, delivery technology expertise in the generics houses has improved to the point where pharmaceutical companies are finding it extremely difficult to gain IP advantage and, instead, need to create real patient benefit and first mover advantage before generics houses actually start coming up with superior products to the original brand.

The traditional branded industry has been slow to really leverage this area to sustain product brands for longer periods. Finding the right balance between NCEs and galenical development in the R&D portfolio is a difficult trade off to achieve. The way to proceed in the future is a change in mind-set. The potential major reasons for development cannot any longer just focus on improving the IP protection for a brand but now must reinvigorate the brand in question. This may be via encouraging depth of usage with current prescribers. For example, offering a sustained release version of an immediate release form, allowing new users to adopt the brand, or creating a children's formulation so that pediatricians find it easier to prescribe. The presentation may improve patient benefits either through a genuine clinical improvement, such as reduced side effects (Sandimmun and Neoral) or, more frequently, by offering improved compliance due to advantageous dosing opportunities, for example, moving from a twice daily formulation to a once daily or from a large solid tablet, that many elderly find difficult to swallow, to melt tablets that allow far easier ingestion.

Figure 8.6 shows the major presentations of selected antiepileptic drugs (AEDs) in 2002. As can be seen, there are quite varied choices within the market, and each presentation offers a slightly modified opportunity for sustaining the brand. The older, first generation brands have the only injectable presentations and as such virtually guarantee their usage in the emergency room (A&E) when a seizing patient is admitted to hospital. Often, patients are later discharged without having seen a neurologist, and, in that case, a follow-on script for the original AED that brought them under control is the most frequent result. Another example is provided by valproate; despite it being an extremely old brand, it enjoys a patent-protected once-a-day formulation in the United States, which provides growth decades after original patent expiry.

Within the European market, where governments are desperate to contain or reduce their health care costs by encouraging generic usage, new presentations are starting to be targeted. The authorities see these new forms of existing brands as a means of limiting generic competition and are starting to act against this trend, irrespective of the potential patient benefit. In the past, many new forms were allowed to be priced slightly above their previous forms within reimbursement systems due to the investment required to bring them to the market. The situation is changing; for example, the French health and social secu-

	Immediate release tabs	CR/OD tabs	Chewables/ melt tabs	Liquids	Sprinkles (granules)	IV
First generation						
Carbamazepine	TDS	BD	Y	Y		
Phenytoin	BD	OD		Y		Y
Valproate	BD	OD		Y		Y
Second generation						
Felbamate	TDS			Y		
Gabapentin	TDS			Y		
Lamotrigine	BD / **OD**		Y		Y	
Oxcarbazepine	BD			Y		
Tiagabine	TDS					
Topiramate	BD		Y		Y	

FIGURE 8.6. Selected AED Brand Presentations. (*Note:* OD = once daily; BD = twice daily; TDS = three times daily; Y (Yes) = presentation available.)

rity minister announced in May 2004 that new galenical forms would be reimbursed at the same level in an effort to control costs.

Additional opportunities to drive patient and physician loyalty exist around packaging one brand with another in a convenient way. The Prevacid NapraPAC was launched in the U.S. market in November 2003, combining the arthritis nonsteroidal anti-inflammatory brand Naprosyn (naproxen) with the hugely successful Prevacid (lansoprazole delayed release capsules). The two brands are supplied as one prescription and one blister pack, the positioning for the new brand being the ability to relieve arthritis pain and reduce the risk of recurrent NSAID-related ulcers.[11]

Clever use of different presentations, how they are protected, and how they meet different physician or patient needs can influence long-term sustainability of a product brand in the market in ways that are often underestimated by the industry.

Product Brand Naming Strategies

In FMCG, there have traditionally been three basic brand name strategies:

- Descriptive brand name (Pampers, Ultra-Bright toothpaste). This name strategy is becoming less frequently used as these brand names are not easy to globalize and are viewed as too generic.
- New brand names (Evian, Perrier, and Ariel). This is a strategy that is being used by many multinationals where it is important that each product brand has a distinctive positioning. A company such as Procter & Gamble exists through its brands and not as a corporate entity. Their strategy is to cover a market with a multibrand approach (Ariel, Dash, Vizir, Bonux, and Dreft, as examples in the detergent market).
- Corporate brand names. In this case, some elements of the name can be linked to the company brand name (Nescafé, Nesquick, and Nestea from Nestlé) or can be fully in line with the corporate name and can serve many different products (BMW, Renault, and Ford) or even product categories. The latter has been long utilized by Japanese multinationals that have, for decades now, given their corporate name to products that belong to different product categories (Honda cars and lawnmowers; Yamaha mo-

torcycles and musical instruments; Canon cameras, printers, and copying machines, etc.).

The growing trend in the consumer world is to use corporate names or big brand names more frequently as "umbrella" brands, as an aid for global recognition. The trend is, therefore, to try and leverage previous brand associations for the new products.

In the pharmaceutical industry, every product brand in fact has two names. The chosen brand name and the international nonproprietary (INN) or generic name. The generic name is present throughout the development process and will be the one used in scientific publications both before and after launch. The generic naming is determined by the World Health Organization (WHO), which publishes the accepted INN as a means of clarification for all chemical entities.

There are at least six separate currently used naming strategies in pharmaceuticals.

Chemically Derived Names

The brand name is based on the scientific name of the molecule. This has been the traditional way of naming pharmaceutical product brands. For example, Cipro for ciprofloxacin, Capoten for captopril, and Risperdal for risperidone.[12] The issue of this strategy is that the product brand name is too close to the generic and might speed generic penetration later in the brand's life. Moreover, it reduces the possibility of finding a unique name that can be used in all international markets.

Therapy Names

The name will be indicative of the disease the product treats (e.g., Procardia for patient suffering from heart problems clearly signals the usage through the "cardia" part of the brand name). This strategy represents a risk as the brand name could also be easily imitated, and generics may find it easy to select a name that is close to the therapy and, therefore, the known pharmaceutical product brand name.

Use- or Indication-Based Naming

The selected name connotes a particular use, indication, or characteristic of a brand; for example, Glucophage, Propulsid, Norvasc,

Ventolin and Cardizem. There is also a risk of imitation from the competition.

Family Name or Drug Class Name

The family name is a brand name that is similar to other products in the same class and is registered by the same company; for example, Mevacor/Zocor, Zoladex/Nolvadex, and Beconase/Vancenase. There is also the possibility of identifying a name that is semidescriptive of a drug class: Tolinase, Micronase, and Orinase.[13]

Corporate Name

The name will contain an identifiable portion of the corporate name tied to a certain product brand or brand line. For example, Sandimmun (Sandoz), Baycol and Glucobay (Bayer), and Novarapid (Novo Nordisk). This strategy is, of course, only powerful when the corporate name is well known and has strong positive associations.

New Invented Name

The name has been created for a specific product; for example, Zocor, Zantac, Zanax, Prozac, Xenical, etc. In the past few years, there has been an overuse of Zs and Xs for first letter but that doesn't seem to dissuade companies or regulatory agencies. The advantage of this strategy is to identify a unique and distinctive name that can also be used for global expansion. It is also potentially easier to protect from a legal point of view.

In reality, these strategies are not very different from the ones existing in the consumer world. We can also identify three basic naming strategies when taking the six current pharmaceutical strategies as a group, i.e., descriptive brand names (linked to molecules, therapy, indication or use, and family or drug product class), corporate brand names, and new product brand names.

FMCG experience, however, suggests that the descriptive names are not ideal for the creation of pharmaceutical brands. They don't offer the freedom to select the right brand name and can just be too close to the INN or generic name, e.g., Capoten and captopril—which one was the brand? Brand names have to be easy to pronounce and, if they are to be memorable, also be short, distinctive, and difficult to imitate. The brand names have to be identified very early in the process as they are part of the brand equity that will be created. New invented names are ideal to meet the criteria of uniqueness and memorability.

Given the trend toward greater use of corporate names as endorsing brands in the consumer world, an association between the corporate brand and product brand name could be a powerful tool in pharmaceuticals. It is, of course, necessary to have already created strong corporate brand names that have a very clear and positive meaning in the mind of consumers. This is far from being currently the case in the pharmaceutical area, following the number of mergers that have occurred over the past fifteen years.

Corporate Strategy

Global corporate strategy has to take account of the global market opportunity, and, at the beginning of the new millennium, this means optimizing your allocation of resources in big markets and big brands that are going to give the highest returns. In reality, over the past decade, there has been a change in attitude to focusing on fewer product brands and success in the U.S. market in particular.

Achieving double-digit corporate growth is no longer the easy task; perhaps it was during the 1990s, and the trend to focus exclusively on potential blockbusters is being pursued with more intelligence nowadays, but it still remains that few brands are selected to receive optimal resourcing. In addition, mergers mean real portfolio management of brands does not as yet really exist within big pharma. It could be argued that portfolio management seems only to be achieved in times of crisis (e.g., during restructuring following corporate failure such as Elan's divestment of its PCP business in the United States). Normally, however, changes are enforced by regulatory authorities, so that competition law isn't breached by the mammoth entities being created.

One of the side effects of this divergence in the size of the opportunity between the U.S. market (transparent, swift registration, free pricing, and small-scale government intervention) and the rest of the world is that when portfolio strategy is defined, the U.S. view tends to drive the major decisions. This was less apparent in the past when Europe represented a significant growth opportunity and Japan was not plagued by biannual price cuts and economic deflation. The key to recent corporate success for non-U.S. companies has been significantly improving their U.S. capabilities and changing the mind-set of the organization. GSK essentially created its decisions-making headquarters in the United States despite being officially a London-based company due to the historical ownership of its share capital in the United Kingdom.

These major factors, therefore, have a significant impact on corporate strategy—at times in a way that is counterproductive to creating a healthy brand portfolio strategy that manages the long term on a global basis as well as the short term. The end of Prozac as a brand was signaled by its U.S. patent expiry and the very rapid penetration of generics, but this undervalues the long-term sales Prozac will continue to deliver to Lilly across a huge number of markets on a global basis.

There is little doubt that corporate strategy in the form of how it allocates its resources has a significant impact on the longevity of a product brand. Portfolio strategy, which not only focuses on new chemical entities, but allows continued clinical development for marketed product brands (both within an indication and in additional indications), will determine brand life to a large part. Attitudes toward funding of landmark studies, where the longer-term view has to be taken to justify the funding or new galenical development has to fit within the overall corporate portfolio strategy.

CONCLUSIONS

The product life cycle (PLC) concept has some applicability to pharmaceuticals but needs to be related to the markets and technologies, and not to individual products. Thinking of a pharmaceutical product brand life, rather than PLC, removes rigidity from the thinking process. Brand life is the length of time an individual brand will remain both a significant sales contributor and an actively managed asset of the company in question.

Life cycle management is a critical success factor for the pharmaceutical industry aiming to achieve maximum brand potential. The traditional view "that nothing can be done about generics anyway" is plain wrong. Strategic branding on a global scale does give many opportunities to extend brand life. Pharmaceutical brand life cycles are often longer (at approximately fifteen years) than many brands from the high-tech consumer industry, an industry that has adopted branding with great success in the recent past.

Brand life determinants include global launch timing, global core market characteristics, product type (biological, prescription, OTC), the indications a brand is used in, its presentations, who uses and how they are used, and, finally, corporate strategy and the financial and intellectual commitment to sustaining them. Interfaces between prescription and OTC business divisions will become increasingly important as major blockbusters expire.

Analysis of global launch timing in the major markets, and their resultant patent and marketing exclusivities, shows that a clear sequence of country launches occurs. Generally, market access is led by the United States, then Germany, and the United Kingdom, followed by Italy, Spain, and France, and finally Japan. Global market characteristics differ widely across major and minor markets. Although regulatory convergence will reduce this variability over time, there will still remain important opportunities for sustaining a pharmaceutical brand over time.

The type of pharmaceutical product brand can dramatically affect long-term sustainability. For prescription brands, there are differences between brands used to treat acute and chronic disease states and whether, for instance, a brand can be taken OTC. The environment will more aggressively encourage OTC switch in the future, and this provides a potential opportunity to utilize strategic brand management to build long-term pharmaceutical brands.

At present, biologicals are largely unthreatened by generics and represent a great opportunity for the industry to build long-term brands. Multiple indications provide opportunities to sustain a brand, a tactic that isn't particularly new but is now being adopted as a classic new-century brand tactic. The development of different presentations of an active ingredient is a classic IP brand tactic normally reserved for aging brands. Nowadays, to achieve longer brand life, the new presentation needs to reinvigorate the brand for the prescriber or the patient.

Sustaining a brand over time is, therefore, possible if enough time and attention is given to regulatory and commercial brand building over an extended period of time. The industry's current practice is too tactical and short-term.

Chapter 9

The Pharmaceutical Industry Brand

WHY DO PEOPLE PREFER TOBACCO COMPANIES TO PHARMACEUTICAL COMPANIES?

The pharmaceutical industry in the early part of the new millennium probably has no better reputation than perhaps the tobacco industry. How could this happen? What went wrong? Were we so busy making money that we lost sight of the future we were creating for ourselves? The media are full of company scandals, such as:

- Defrauding Medicare
- Promoting products outside their registered indications
- Making gargantuan profits while sick people can't afford the basic drugs to stay alive
- Denying the third world the AIDS drugs it needs to avert a population disaster of untold proportions

Whether we like it or not, the debate around the world, even in the United States, is pharmaceutical profit versus public good. At a time when health care systems around the world are creaking with an aging population, spiraling costs, and increased consumer expectation, the traditional industry profit margin is under attack. No other industry has the same opportunity to do good, no other industry has the opportunity in the future to deliver more. Eradication of previously lethal diseases via vaccination is now accepted as normal; pharmaceuticals reduce the cardiovascular impact of our twenty-first-century life styles; and we are starting to make progress in the treat-

Pharmaceuticals—Where's the Brand Logic?
Published by The Haworth Press, Inc., 2007. All rights reserved.
doi:10.1300/5836_09

ment of depression and other previously poorly understood psychiatric diseases, offering the hope of a normal life to people with countless conditions and diseases.

During the last decade, it became increasingly apparent that brands can exist at country, regional, and industry levels, as well as the more easily accepted company and product levels. The pharmaceutical industry brand is bankrupt. Frankly, we've done a lousy job of managing our future—why? Perhaps we didn't understand that someone has to manage the image, the brand, of the industry itself. The pharmaceutical world is made up of diverse players that include those from the medical professions, patients (or consumers), the distributors of drugs, and most important, the insurance companies and governments who pay. It is a peculiarly political industry because how it makes money is heavily influenced by the political decisions of governments (or insurance companies in countries where direct government intervention is less common).

If we briefly look at two other industries, those of the tobacco and oil (gasoline) producers, which by rights should have a worse brand image, we find a contrast. Their images are surprisingly good, and that's probably because they have been actively managing their brands for a long time.

Tobacco and oil give pleasure! Pharmaceuticals are negative goods; they reduce or alleviate, or occasionally cure, the negative effects of a medical condition. The healthiest person in the world wouldn't need our products, and we've forgotten this fundamental. In the 1970s, 1980s, and the early part of the 1990s, we got away with what we did, largely due to having a generally poorly informed general public who wanted to place their trust in the physician and the people who gave them their wonder drugs. AIDS and the Internet saw to the end of that—suddenly, there was a new disease that tended to affect a small but well-educated and highly motivated section of the population, a group of patients who became experts about their condition. They no longer respected the physician to make those life or death decisions that nobody had wanted to be involved with in the past. There was also a new medium of communication that started to quickly educate the consumer about their drugs and the iatrogenic effects (or unwanted drug-created side effects) caused by the drugs that were supposed to help.

The tobacco industry has meticulously constructed a positive brand image over decades—associating smoking with sophistication, self expression, and success, and being able to enjoy the good things in life. Everyone knows nowadays that smoking kills, but the addictive nature of the product and the equally addictive strength of the brand have meant huge swathes of the population happily ignore the facts.

The oil or gasoline industry is another good example of a brand that has been built over decades. Gasoline leads to thrills—everyone around the world loves their cars, and gasoline is what allows them to indulge in this pleasure. The extraction of oil may also be causing an environmental disaster; abuse of human rights, and untold negatives effects on the ozone layer, but, generally speaking, working in the oil industry is a worthy profession, blighted at times by stories of abuse in Africa, and under rightfully greater scrutiny, but all the same it is generally worthy.

A critical incident for the pharmaceutical industry occurred in early 2001 when news media covered a breaking story that the pharmaceutical industry in South Africa was prepared to take court action to defend patent rights for AIDS drugs. Within days, this news story had circled the globe, gathering pace as it passed through the time zones, turning into a global debacle for the pharmaceutical industry, damaging the industry brand around the world.

The end result of this incident (and various others) was a political triumph for health activists at the World Trade Organization (WTO) ministerial conference in Qatar in November 2001. While the WTO's TRIPS (trade-related aspects of intellectual property rights) agreement wasn't changed, it allowed developing countries to take necessary actions to protect public health, such as issuing compulsory licenses to import cheap generic copies of HIV/AIDS, malaria, and tuberculosis (TB) medicines.

There is little doubt that the day the big pharma companies decided to sue in South Africa to protect their legitimate patent rights on AIDS drugs, it created a stick with which to beat the industry for decades to come. Irrespective of whether or not the industry was strictly correct, or how vehemently you believe that large profits are required to fund the escalating cost of R&D, an average member of the public was never going to see anything other than a blatant attempt by a rich industry to protect its profit margin in a country that desperately needed help (and still does), a classic own goal that is inexcusable in a global industry. This issue is unlikely to easily go away if the Medecins Sans Frontieres (MSF) report "Drug patents under the spotlight: sharing practical knowledge about pharmaceutical patents"[1] is anything to go by. One of the many claims made in this report was that the industry was being lazy in its arguments, suggesting that without patents there will be no new medicines. MSF argued that as Africa accounted for just 1 percent of global sales, that if no sales were made at all on the continent, the industry's profits would be affected negligibly. This kind of easy retort shows the industry has not been giving these issues sufficient thought.

VICTIMS OR CREATORS OF OUR OWN DESTINY?

This question is worth evaluation—are we victims or creators of own destiny? Well, it could be argued we are where we are, as a direct result of poor communication around what we do.

The industry hasn't really taken the care to evaluate how best to communicate to a consumer audience of the good we do. Do we cure people, or at least alleviate their suffering, rather than create side effects that are worse than the original disease? If we have to communicate difficult issues to the medical communities to maximize the opportunity for our blockbuster brands, then the market research is done, the right structural changes are made to facilitate that communication, and we act as marketing professionals.

Why haven't we done the same for communication with the ultimate end users of our brands—the patients and the payers? Are pharma marketing personnel and senior management so isolated from the real world by corporate internal activities that they can't see the changes going on around them? Merger activity could be one causal factor as they suck huge amounts of energy out of the organization, and the focus is not external, not about adapting to the changing external environment, but about surviving the internal politics and cost cutting. The elimination of overlapping resources (people or projects) tends to be the focus for success. Many mergers, however, are not successful in that the companies in question have seen a reduction in their combined market shares as a result of the merger—perhaps because downsizing doesn't encourage innovation in an industry peculiarly dependant on its personnel. Due to the prediction that companies that spend less than $5 billion on R&D will not be able to keep pace with the rising costs of R&D in the future, Decision Resources[2] thinks further structural consolidation by the industry will create only fifteen to twenty innovative companies going forward. Under that scenario of continued consolidation, inward focused senior management will continue to be the norm.

Trying to maintain historical rates of growth, in the double digits, has perhaps created a cruel and vicious industry. Global executives, who spend so much time traveling away from home that normality is a difficult concept to appreciate, are exactly the ones who need to be keeping themselves aware of global changes in consumer sentiment to the industry in which they work. One top-ten big pharma company had well-advertised core values in the 1990s that could be seen when visiting its offices. Those value statements included such terms as integrity, team work, performance, etc. The reality, however, was that staffers were reported to joke, unfortunately all too seriously, that the core values were, in fact, performance, performance, performance, performance, performance, performance, and performance. If that is how the people who work in the industry think, what is everyone else going to conclude?

As the pipelines were drying up and double-digit sales increases became more difficult to achieve, the industry turned to aggressive patent term defense as a legitimate critical success factor. Sometimes these tactics went wrong, as when BMS eventually agreed to pay $670 million to settle the majority of the antitrust claims surrounding BuSpar (buspirone) and Taxol (paclitaxel).[3] The content of the Taxol case was interesting because BMS had obtained the license to the product from the U.S. government. The lawsuit asserted that BMS had stated during initial negotiations that obtaining a patent on Taxol was not possible and therefore generics would be available by 1998 following a five year marketing exclusivity award. The suit contended that BMS fraudulently obtained U.S. patents and used them to block generic competition for more than two years. These events led to a series of publicity disasters that tarnished the company's reputation and obscured the innovative nature of the oncology "Bristol" company. At the time, Bristol-Myers Squibb was the de facto "oncology company," bringing hope to millions of cancer sufferers around the world and had managed to create a real and powerful franchise brand through a combination of their own products and excellent in-licensing from the outside.

Thinking again about the subject of profits—we have to make those profits to be able to invest so heavily in R&D, right?—we are probably one of the biggest and most profitable industries in the world. We spend more on R&D than virtually any other industry sector. But R&D productivity has dropped over the last twenty years, in part due to the ever-increasing regulatory hurdles required to bring a drug candidate to market nowadays. There are well-renowned third-party groups such as the Center for the Study of Drug Development at Tufts (CSDD), which provide data on the escalation in costs associated with performing clinical trials, but what we haven't done is leverage those independent data sources to work for us. Do we, in reality, talk widely about the impact this has on our business, of how this drives our need to produce double-digit sales growth to fund those increasing costs? Do we talk with suitable understated pride about the impact vaccination has had globally on disease and life expectancy? Do we segment consumer audiences to provide them with appropriate targeted messages about the good we do? Unfortunately, the answer is no, but what we appear to be able to do as individuals is talk with pride in industry journals about how we are working to help cure people of disease. The problem is that often this is in a way that looks a bit desperate, as we try to justify some of the unsavory commercial tactics we appear to use.

One of the cardinal sins of consumer marketers is to allow someone else to destroy your brand. Years and years of careful work can be destroyed in days, hours, or minutes, thanks to bad publicity, something ably demonstrated by the stratospheric fall of Martha Stewart. This icon of modern American living and head of MSO was caught up in the Imclone scandal surrounding the Food and Drug Administration's (FDAs) refusal to accept Imclone's New Drug Application (NDA) for the anticancer drug, Erbitux (cetuximab). The story goes that following a $2 billion deal with Bristol-Myers Squibb covering Erbitux in September 2001 and a subsequent FDA meeting in December where the FDA expressed concern about the filing, Sam Waksal, the CEO, tipped off family members and friends in advance of the problem being made public on December 28, 2001. Waksal's friend, Martha Stewart, sold her shares on December 27, and the subsequent investigation did untold damage to her hard-won reputation and, therefore, directly affected the results of her company. The company depended on her brand image to market numerous products, including the *Martha Stewart Living* TV show. Other notable failures include the failure of Arthur Andersen, the audit company, to survive the Enron scandal. Interestingly, their management consultancy arm did survive thanks to a clever corporate brand-building campaign in advance of the "shredding" allegations that proved so costly.

The problem is that currently the pharmaceutical industry appears immune, and more worryingly acts like it is immune, to this type of brand damage. There are many current pieces of bad publicity that just tend to roll on unmanaged, e.g., regular fines paid by top players for allegedly defrauding Medicare in the United States; the potential cover up, according to former New York Attorney General Eliot Spitzer, of serious side effects of Paxil/Seroxat and Celexa/Lexapro in children and/or adolescents; whistleblower cases such as that highlighting off-label promotion of Neurontin in the days of its Warner Lambert ownership; regular price increases in the United States while discounting to Medicare, therefore devaluing the value proposition and hiking the price to those having to fund themselves; your-money-or-your-life pricing for new innovative brands, particularly in the oncology and cardiovascular areas; and the continued low access to important medicines in the developing world.

The industry is well known for having more than 600 full-time political lobbyists in the United States alone, and no doubt they are busy clearing up the seemingly endless mess we seem to be able to get ourselves into. Industry associations exist, but we have to question whether they are effective? Are the people who sit on them the wrong kind of people to try and resolve major public image questions and protect the pharmaceutical industry brand? As an example, the EFPIA (European Federation of Pharmaceutical Industries & Associations) generally only agrees on one thing—they hate parallel trade in Europe.

The industry may well be at an inflexion point of how it is perceived by the general public. In recent unpublished research, it appears that, in Europe, public opinion on the industry is not destroyed—yet! The public currently blames government health care policy and not the industry for inequalities and poor general health provision. So, there is still perhaps time to act before the job of brand revival becomes impossible. The use of heavy direct-to-consumer advertising (DTCa) in the American market may have accelerated the decline in industry image there. An NBC/*Wall Street Journal* survey[4] asked which industry had stopped listening and being sensitive to its customers: 34 percent named pharmaceuticals. This was only slightly mitigated by the fact that health care (insurance systems) fared worse at 36 percent.

Coming back to the original question, are we victims or creators of our own destiny, our chosen roles being to create new medicines and to find new ways of using them or delivering them. Whether we like it or not, the role of the other players is to decide how successful commercially we will be, in light of the fact that health care absorbs between 6 and 14 percent of gross domestic product (GDP) in the developed world. Governments decide which brands are registered, how they may be sold (OTC versus Rx), how they are marketed, and what information can be made available to health care professionals and consumers (DTCa). In the future, this control is likely to further increase as once independent physicians see their clinical freedom increasingly eroded by governments and insurance companies alike. In the past, it appears, we have been the major contributors to the mess we are in, especially as we know we are a political industry. Unfortunately, it is less clear whether we will be able to get ourselves out of that mess in the future. The pharmaceutical industry brand has become so tarnished that it is going to take many years of careful brand management to bring it back to the position it deserves.

Again, back to another central question at the beginning of this chapter—why do people prefer tobacco companies to pharmaceuticals? Well, for starters, they acted as one. The tobacco industry was and still is heavily active in political lobbying, as is the pharmaceutical industry. They were represented at the highest level during the 1980s. In private discussions with a member of one of the European Commissions Director General (DG) departments in the European Commission in the early 1990s, someone commented that when Margaret Thatcher phoned the relevant commissioner, then invariably she, and the message she was delivering, got through. Lady Thatcher, the ex-prime minister of the United Kingdom, was reported to be active throughout her premiership in supporting the tobacco lobby, something for which she was prepared to take the negative press for.

With the admission that tobacco products are causing premature death has come new tactics from the industry focused on increasing corporate responsibility. British American Tobacco (BAT) has launched a social report and has tried to enter into dialogue with stakeholders, its key brands being Dunhill and Rothmans. Many in the antismoking lobby, such as ASH (action on smoking and health) will not enter dialogue, due to fears that dialogue will give respectability. Another such tactic is the high-profile campaign funded by Philip Morris and Japan Tobacco International, which is an international ad campaign placed on MTV that aims to prevent kids from smoking.

The activists suggest that the tobacco companies are using corporate social responsibility as a smoke screen for its activities, and that with the European advertising ban making some impact, they will just redirect their spend to the developing world, but still the industry brand appears to be making progress.

We are, therefore, both victims and creators of our own destiny. But if an industry such as tobacco can make progress, then the pharmaceutical industry can do far better!

WHAT THE FUTURE BRINGS

The future for the industry brand looks bleak. As the twenty-first century really gets underway, activist groups are becoming more effective in their organization and communication. Demonization of pharmaceutical companies, and therefore its associated industry brand, will continue. The focus will change over time; for instance, we've seen a move toward attacking the industry over its marketing practices (particular in the United States and France) after many years of activism on patent law abuse as a means of delaying the entry of generics.

The realization that price controls in the U.S. market are a possibility in the future could have a seismic effect on the industry over the next decade. Post 1990s, realism meant that all the top players had to focus on the rapidly growing U.S. market to remain competitive, and drive continued top-line sales growth as other major markets, such as the EU and Japan, were becoming more difficult.

There are many ways to manage prices indirectly, without taking the political heat, and those mechanisms will become increasingly attractive in future decades. Estimates provided by Burstall and Reuben[5] (see Figure 9.1) suggest that during 2001 the U.S. market accounted for $168 billion of the global $376 billion annual sales (45 percent), but, more importantly, delivered $70 billion in estimated profits or 84 percent of total world's profit. This U.S. activity equated to a margin greater than 40 percent. The next biggest contributor was Japan with $48 billion sales, $18 billion profit, and a margin of 37 percent.

Activism by American seniors in particular has meant an increased awareness of rising drug costs and brought huge media attention onto the parallel trade between Canada and the United States. The industry has responded, and been supported by the FDA, by arguing that the supply chain cannot be guaranteed, and that the trade opens the door to potential counterfeit or substandard products reaching American consumers. Irrespective of short-term success in controlling what in

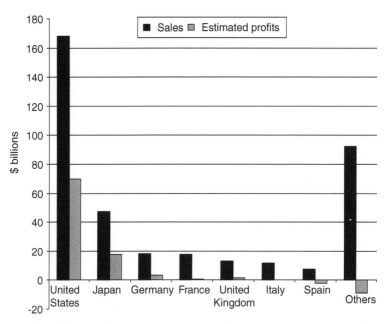

FIGURE 9.1. Estimated 2001 USD sales and profit by country. (*Source:* Adapted from Burstall M. L and Reuben B.G at Decision Resources Inc. "The Outlook for the World Pharmaceutical Industry to 2015.")

reality is a tiny proportion of the overall market; the fact that U.S. citizens pay significantly more for drugs is now common knowledge after being highlighted by many, including the ex-FDA commissioner Mark McClellan. The political heat being created by access to affordable medicines in the United States raises the prospect, at the very least, of price controls. The merest hint of its likelihood would potentially have a devastating effect on global pharmaceutical share prices and could single handedly rebase the financial prospects for the industry in the future.

The European and Japanese markets are unlikely to get any easier. Although Japan is now starting to move out of its long economic deflation phase, regular price cuts remain a key tool for control of spending. The European market continues to be difficult.

Many countries already maintain ever increasing control on the reimbursement of medicines and are actively encouraging generics. At the same time, they are becoming more successful in establishing cost shifting to the European consumer via lower reimbursement rates. There is the specter of a potential Euro NICE (National Institute of Clinical Excellence) to help restrict prescribing and make rationing decisions as well as legitimizing slower access to innovative medicines.

In an analysis of strategies to reduce spending on prescription medicines by Decision Resources[6] (Figure 9.2), it can be seen that cutting drug prices is one of the actions that gives instant results (along with cutting reimbursement lists, transferring prescription products to OTC, and raising copayments). In addition, cutting prices requires little political effort (because the pharmaceutical industry is an easy target) and is acceptable to both physicians and patients with only the pharmacists disappointed. It is hardly surprising that governments favor this activity while they work on the more long-term solutions of encouraging generics, restricting prescribing, reforming distribution systems, and allowing pharmacists to substitute.

The European Medicines Agency (EMEA) has at times appeared to be a political body, despite its function being to evaluate new chemical entities on a scientific basis. Its closed-door deliberations are, however, reported to be at times more influenced by budget concerns than science when it comes down to approving innovative new medicines. One late 1990s example being the scarcity of clinical data available to

Strategy	Savings	Effect felt	Duration of effect	Political effort needed	Acceptable to		
					Doctors	Patients	Pharmacists
Cut prices	Medium	Instantly	1-2 years	Small	Yes	Yes	Usually not
Cut reimbursement list	Medium	Instantly	Uncertain	Large	No	No	No
Transfer Rx to OTC	Uncertain	Instantly	Many years	Small	Maybe	Yes	Yes
Raise co-payments	Medium	Instantly	Uncertain	Large	No	No	No
Encourage generics	Large	Slowly	Many years	Moderate	Increasingly	Maybe	Maybe
Restrict prescribing	Uncertain	Slowly	Many years	Large	No	No	No
Reform distribution system	Medium	Slowly	Many years	Large	Yes	Maybe	No
Allow Pharmacist to substitute	Uncertain	Slowly	Many years	Moderate	No	Increasingly	Yes

FIGURE 9.2. Strategies to reduce spending on prescription medicines. (*Source:* Adapted from Burstall M. L. and Reuben G.G. at Decision Resources Inc. "The Outlook for the World Pharmaceutical industry to 2015.")

justify the narrow indications approved for the second two glitazones to market (Takeda's Actos [pioglitazone] and GSK's Avandia [rosiglitazone]). A huge swath of patient data was ignored in favor of a narrow indication where total patient numbers experience numbered less than 100 in a clinical trials program of multiple thousands. The outcome appeared to be a budget decision, with science being relegated to a minor role.

Accession of the Central and Eastern European (CEE) member state countries to join the original fifteen E.U. member states in mid-2004 will bring more dynamism to the European market. The jury is currently out about the real impact on patent protection, pricing, and generics because of the various mechanisms to manage the change, but the market is unlikely to become easier.

Added to these market uncertainties is the unprecedented patent expiry situation over the coming decade. Big pharmaceutical companies created huge blockbuster brands during the 1990s and earlier this century. As a direct consequence of this previous success, replacing those lost sales is going to be very difficult.

The traditional industry tactic of aggressive patent defense is being countered by confident generic manufacturers who are prepared to take calculated courtroom risks to secure exclusivity claims for their first-to-market generics. To compound all these problems, the industry is suffering from a dry pipeline until genomics really starts paying off.

We have a few opportunities on our side—counterfeiting of pharmaceutical products and the potential problems associated with inferior-quality products has now been raised in the minds of the consumer. Numerous examples have been given widespread international coverage, especially those in China, the world leaders in copying branded products. Chinese patients have suffered as a direct result of taking medicines that were either ineffective or plain dangerous. This was a consequence of as much as 15 percent of all Chinese medicines now being counterfeit according to some estimates.[7] The Latvian medicines agency took action in May 2004 to withdraw the marketing authorizations for 398 pharmaceuticals that did not comply with good manufacturing practice (GMP) standards, an action that hit both local Latvian firms as well as companies from other Baltic States and Russia.[8]

Continued downward pressure on government expenditure could go either way for the industry, being negative if drug expenditure alone is targeted and positive if the objective instead is optimizing healthcare as a whole. This must be a key focus for the future if the industry is going to work on the industry brand and the value we deliver.

MOVING FORWARD

So, what would be the arguments the industry could use to rebuild its status and look after the long-term brand? At this stage in history, with a burgeoning aging population in the United States, Japan, and Europe, rationing of health care is a necessity. Rationing or triage has always been a fact in society; it is just that, generally speaking, society hasn't been ready to discuss the subject in a coherent fashion. Analysis of the cost of medicines (within the bigger picture of the cost of health care provision) reduces the impact of drug costs to approximately 10 percent of the total. Health care systems will probably change in an incremental manner, due to their political sensitivity, rather than by revolution. They have widespread support within their cultures, especially when taken in the context of the amount of investment already made over the years.

The pharmaceutical industry been an easy target as this discreet health care cost is an easily identifiable budget, which has grown year on year as advances have been made in medicine. A carefully constructed public relations campaign (some may be necessary to get the facts across) coordinated on a global level could start to redress the balance of the political situation. One thing is guaranteed: health care is a political hot potato, and everyone likes to blame an industry that makes profits. Individual companies can have a minor impact. GSK, in particular, has been aggressive in advertising the good side of the industry during 2004.

Pharmaceuticals is a global business, which employs hundreds of thousands of people directly and indirectly—why not use the power of those individuals within their communities on a global scale? Again, the real power would come from coordination around the world, a lesson learned while marketing blockbusters during the 1990s.

One of the problems with the current industry associations is that the people who work in them are experienced professionals often in their fifties. With a comfortable retirement in immediate prospect, they are highly unlikely to challenge how they achieved their success, when planning for the future. Consumer companies trust their brands to a younger generation. We should ask ourselves what our health care providers want—lower costs, controllable costs, predictable costs, and improved outcomes for the patient. So, if we were consumer marketers in charge of the industry brand, what would be our value argument? What areas of potential value would we need to research with consumers, health care providers, and politicians to hone our brand messages?

- Acknowledged benefit of drug treatment over surgery, medical devices, or psychotherapy
- Patients being able to receive care at home
- Reductions in associated health care costs of nursing or supportive care
- Reductions in risk profile for patients at high genetic risk of certain types of diseases
- Increased understanding of consumer health issues as a result of industry activity
- Specific gene-targeted therapies that are highly effective in small patient segments

Collection of data to support positive messages and arguments will have to be done carefully. Pharmacoeconomics data presenting drug treatment compared to other forms of therapy could be very advantageous, as it would focus the debate on the best way to provide care.

Drug treatment tends to dominate in treatment of infections, asthma, skin complaints, gastric ulcers, and, increasingly, diabetes and cardiovascular disease. Major advances are being made in many central nervous system (CNS)-related diseases and, perhaps less important, but a major potential PR boon, lifestyle problems.

If the industry started to focus its attention on looking after the long-term future of the industry brand, showed some intellectual interest in the issues, and invested a small amount of expertise and money in execution, a huge difference could be made. We are not yet past an inflexion point where public awareness and understanding of the value that this industry brings will forever be negative.

CONCLUSIONS

The pharmaceutical industry brand is bankrupt, less popular perhaps even than those of the oil (gasoline) and tobacco industries. We operate in a political environment and a lack of management of the industry brand has led to a devaluing of our societal value.

The debate is often pharmaceutical profit versus public good. Medicines are negative products (they don't give pleasure). The oil and tobacco industries have worked hard at managing their brands. The pharmaceutical industry brand is currently unmanaged.

We are both victims and creators of our own destiny. Mergers have contributed to the internal focus of senior management within the industry; continued consolidation is forecast to leave just fifteen to twenty innovative companies conducting top-class R&D in the future. The process will probably facilitate further introspection, rather than outward-looking industry brand building.

The pharmaceutical industry is a major contributor to its brand image problems. Over-robust patent defense, questionable marketing tactics, and accusations of denial of drug access to the third world shows a lack of care being taken of the future. This allows governments to shift the blame for spiraling health care costs onto our shoulders rather than having to deal with a public debate on rationing and the best way to provide health care. The industry could be at an inflexion point for consumer credibility and does not know it.

The future outlook for the industry is bleak. Difficult market conditions, patent expiries, and continued political focus on controlling our activities are expected. The industry is in danger of trying to manage the future using the successful tools of the past.

The health care markets within which we operate will see evolutionary change rather than revolutionary. A crisis could, however, be created by intensifying the need to rationalize health care spending, with price control being a key potential risk. Continued downward pressure on government expenditure could go either way for the industry, being negative if the focus is drug expenditure alone and positive if the objective instead is optimizing provision of health care as a whole.

A coordinated global campaign, harnessing well-conducted research into how governments and the key consumer constituents perceive the issues is needed while we still have time. This would allow targeting of appropriate messages about the good the industry delivers. This would start to redress the balance of public opinion after years of neglect and, over time, rebuild the tarnished pharmaceutical industry brand.

Chapter 10

The Pharmaceutical Business Model

At its simplest, the business model that a big pharma company follows is one of product creation. The elements of that creation are research and development (R&D) and promotion by massive sales forces, followed by aggressive patent defense. R&D is required to create patentable products that offer cures or significant amelioration of symptoms for a disease and then try to extend the longevity of sales by life cycle management once a product brand starts to generate revenue. Massive sales and marketing organizations try to minimize the time to peak sales through huge commercial spend, and intellectual property departments work extremely hard at delaying generic competition to their branded patent protected medicines.

Obviously, there are vital subprocesses to this simplistic model, such as achievement of the best price while the product is patent protected, the assessment and acquisition of the required critical mass within the global organization, the role of business development and licensing within both R&D and sales and marketing (S&M), and the targeting of therapeutic areas that allow significant future growth potential, but the basics of the business are easily described. The acceptance of brand destruction is implicit in the model, the time horizon of commercialization is short, and after ten to fifteen years of significant investment the products are cast out to become cash cows, often without much thought. The process is therefore one of product creation not brand creation.

In contrast, the FMCG sector, in particular, concentrates on creating the brand. It still does its R&D, applies for patents, and sells the products, but the whole organizational focus is on the long-term

Pharmaceuticals—Where's the Brand Logic?
Published by The Haworth Press, Inc., 2007. All rights reserved.
doi:10.1300/5836_10

value derived from creating the brand from the product. Figure 10.1 illustrates some of these differences.

The FMCG business has a longer time frame for recouping its investment, in the order of decades rather than years. To some players, the life cycle doesn't need to exist if the product brand is always kept current and up to date. Investment, therefore, maintains the brand over time and, rather than product brand destruction, portfolio rationalization ensures each product brand plays a role within the portfolio. In effect, the brand is the asset, and the focus is one of brand creation.

So far, the simple pharmaceutical business model has been sufficient for the industry to achieve huge success over an approximate fifty-year period. Over the last two decades, the blockbuster product model in particular has held the attention of senior management around the industry due to its track record of success irrespective of the size of the company. The rationale for this being that just one blockbuster can make a company in the long run, e.g., the proceeds from Zantac made GSK what it is today, allowing it the financial clout to acquire other businesses and sufficient funds to just about keep a pipeline alive. There have been really only two subtleties, either a global mass-market blockbuster business model, as typified by Lipitor (atorvastatin from Pfizer), or a globally targeted blockbuster

FMCG	Pharmaceuticals
A brand creation focus	A product creation focus
• Time horizon long (decades) • Life cycle doesn't need to exist • Huge marketing effort to create the brand • Investment maintains the brand over time • Brand destruction does not exist, portfolio rationalization does THE BRAND IS THE ASSET Strategic and brand management vital	• Time horizon short (years) • Life cycle management necessary • Huge R&D effort to create the product • Patent expiry signals loss of resources • Product cast out to provide revenue for new products THE PRODUCT IS THE ASSET R&D and sales managment vital

FIGURE 10.1. Differences in attitude to brand building.

model, as seen with Gleevec (imatinib from Novartis). Naturally, there are numerous examples of peculiarly local products becoming best sellers (particularly in Japan), but this phenomenon hasn't driven the overall growth within the pharmaceutical market over the last two decades.

According to brand commentators, such as Simon and Kotler,[1] the evolution of the market to more targeted drugs will arrive quickly with a brand like Lipitor (currently catering to the mass market) evolving into Lipitor (a), Lipitor (b), and Lipitor(c), each individually targeting specific genotypes, and this by only 2015 or 2020. They also predict that from the thirty biologics creating $15 billion sales in 2002 that this could become 100 biologics on the market by 2012, with up to $50 billion in sales. With the biologics producing more than $35 billion in 2003, their sales prediction looks safe. In fact, it looks very conservative, considering 25 percent of the industry's 2003 pipeline was biotech. The rapid move to highly targeted drugs across the broad spectrum of disease is less likely, which is an evolution that could take thirty to forty years instead.

THE BLOCKBUSTER MODEL: WHAT HAVE WE LEARNED?

The common themes to the model can be oversimplified by the phrase "shape the product, shape the market, shape the company," a phrase perhaps attributable to McKinsey.

One of the major advantages to adoption of the model is simplification of the clinical profile on a global basis within the context of shaping the product. One global positioning leads to faster (and sometimes simultaneous) regulatory approval around the world, each local data sheet or prescribing information having the vast majority of wording identical to its neighbors. The world is smaller, and health care systems are put under similar pressures, causing them to demand more evidence from the industry. At the same time, physicians are traveling more to international medical congresses, and, as a result, differences in treatment protocols are diminishing; they still exist and are widest at the level of primary care physician (PCP), but, in the specialist areas, thinking is converging, and that thinking is driven by the data. Companies can shape the product and also the market with

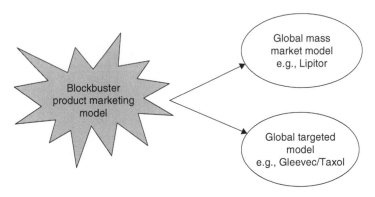

FIGURE 10.2. Two major variations to the blockbuster model exists.

their clinical trial strategy and subsequent data analysis. This ensures that key opinion leaders (KOLs) are involved in major clinical studies, creating advocates for the product brand in question.

Later, when the brand is more mature, a tight control on the available data and the ongoing clinical studies continue to pay dividends. Gone are the days when every sizable local affiliate could do its own market-specific clinical studies, allowing subtle local prescribing information changes, and potentially open up the long-term safety record of the product via poorly informed guesswork.

Shaping the organization is also vital to implementing the blockbuster model to its optimum. Pfizer, for instance, realized earlier than anybody else that when it came to in-licensing having the biggest and best-rated sales force paid off at the negotiating table. To become partner of choice in primary care gave Pfizer a differential advantage when it came to marketing products in highly competitive therapy areas. Other companies found out the hard way that when it came to a specialist brand, in a highly technical area, the organization that had made them successful in primary care no longer had the right skills for targeted technical promotion.

The two variations of the blockbuster model are shown in Figure 10.2.

Global Targeted Model

Numerous brands are good examples for this particular model, the ones that come immediately to mind being Taxol (paclitaxel from

BMS), Gleevec from Novartis, and a number of AIDS drugs. These types of product brands tend to target a disease area of high unmet medical need, often with quite a small but well-informed patient group, cared for by a limited number of specialist physicians who maintain a global network of interaction. These product brands are driven by a positioning built from clinical trails results, and it is no surprise that many of the obvious examples come from areas of high societal interest—modern society finds cancer and AIDS as justifiably emotive subjects (oncology is currently the most researched therapy area in terms of numbers of active compounds in the industry R&D pipeline). Central to success for any company entering these areas for the first time is establishing credibility with highly specialized stakeholders and a buy-in to the activities of the company. The interaction between marketing and medical affairs is particularly crucial, as often significant off-label sales can be generated by physician interest in alternative disease states to the registered indications. These specialist groups can be effectively covered by comparatively small but highly trained sales forces and medical science liaison personnel. In addition, global communication networks can be quickly established through international medical congresses, global advisory boards, and the Internet rather than mass sales force tactics. An effective sales force in the United States may be only 100-150 reps, while 2,000-3,000 would be required for a primary care product. In the United Kingdom, the number would be ten to twelve, while approximately 180 are needed for primary care physicians.

Global Mass Market Model

The best and obviously most successful example for the mass market model is Lipitor (atorvastatin from Pfizer). Warner Lambert originally held the rights for atorvastatin, a product with a breakthrough clinical profile even for the then new statin class. It was more effective at lower doses than its competitors and appeared to have greater effect on triglycerides than the four other marketed HMG-CoA reductase inhibitors. Pfizer had long recognized that sales force size was of critical importance to licensing partners and had already built, through mergers, one of the biggest global sales forces in the industry. They then aggressively bid for Lipitor copromotion rights, beating off a number of other companies to successfully partner with Warner Lambert.

The two companies married the excellent clinical profile of Lipitor to huge prelaunch spending, concentrating on comparative clinical studies. This was allied to rapid registration globally, strong manufacturing capability, and massive sales force resourcing before and at launch. After a short physician detailing period, they went to the patients via direct-to-consumer advertising (DTCa)—sales take off was rapid. The rest is history to some extent, when it became clear that Pfizer might lose the future potential growth provided by Lipitor, they launched a hostile takeover of Warner Lambert to ensure the asset was secured. The key weaknesses of the product, being fifth to market and lacking in outcomes data (morbidity/mortality), were overcome by aggressive activity and spending in a true mass market blockbuster case study.

Will the Blockbuster Product Model Survive?

The blockbuster model has dominated the strategic thinking of the pharmaceutical industry over the last two decades. There are, however, many factors that are starting to point to the end of this highly effective model. This means that, in the future, many other aspects are going to have to be taken into account to decide successful strategy. Although there will still be a few players who find success with the blockbuster model merely through quality of their implementation (shaping the product, the market, and the company), for the majority of the rest they will have to find new value. Companies need to recognize that the winners of the future have to differentiate themselves and their products as unique for physicians, payers, patients (consumers), and investors.

Physicians are getting more difficult to influence through traditional rep activity. Success in the late 1990s was driven by increasing the number of reps every year (often by an annual rate of 20 percent or more), but now we appear to have reached saturation. The physicians wish to spend less time seeing multiple reps detailing the same brands and are introducing quotas for the number of reps they will see from a particular company. Would you really want to see 12-15 GSK reps if you were a primary care physician in the United States or nine in the United Kingdom? At the same time, physicians' clinical freedom is being reduced due to formularies implemented in a myriad of different ways around the world by governments and/or insurance companies. In the United States, in the early 1990s, about 60 percent of all medicines were paid for by individuals, but a decade later 70 percent are now paid by Health Maintenance Organizations (HMOs) and insurance companies, and, unsurprisingly, they want to have a bigger say in what patients get prescribed. In the other major markets around the world, the government still largely pays the bill, but they are implementing numerous ways of "managing" prescribing such as the introduction of the "National Institute for Clinical Excellence" (NICE) in the United Kingdom, or via reductions in reimbursement rates for expensive patent-protected branded products.

R&D is getting tougher, and, with many of the "low-hanging fruits" having already been picked, the industry has to move to targeted discovery and look to biotech for its pipeline. Big pharma is more set up for mass market lead screening than for molecules targeting individualized specialist populations, the successful registrations of new chemical entities (NCEs) have been dropping in number since 1999. At the same time, conducting clinical trials is becoming more difficult (and more expensive), finding quality investigators is a challenge, and recruiting willing patients more troublesome in a world that has seen the withdrawal of both Lipobay (Bayer) and Vioxx (MSD) in recent years. The average cost for a launched drug is now estimated to be as high as $800 million according to the research conducted by Tufts University.[2]

Big pharma, in particular, has to be able to reconcile critical mass, i.e., the huge size many organizations have become and the impact this has on agility, and the ability to make decisions and implement them. The top organizations are huge, with many layers of management and a high internal reporting workload that appears to be less productive, particularly in R&D, than smaller organizational structures. Sales forces are huge by any industry's standards and, as a result, become the major P&L cost item for the company; marketing is often then relegated to the function of "feeding" this huge manpower investment.

Another aspect related to size is the need for portfolio management. Increased size has been driven by the heavy consolidation that has characterized sector behavior during the 1990s. This has continued in this century, with numerous mergers, and now hostile take-overs such as that between Sanofi-Synthelabo and Aventis to form Sanofi-Aventis. Portfolio management appears to be alive and well in the R&D world, where there is at least discussion of the trade offs required to create a portfolio of compounds that mitigates opportunity, risk, and the probability of success. The same can't be said of the massive unrationalized sales portfolios that have been created by these mergers—the only strategic thinking that appears evident is the ability to value cash cows on a ten-year net present value (NPV) with the addition of a terminal value (i.e., no financial risks are taken). This leaves the vast majority of big pharma portfolios broad, fat, and underpromoted, despite the sales force being such a huge proportion of costs. In-licensing and external product acquisition merely add more complexity and trade offs into the sales force resource allocation problem. No really innovative thinking has yet gone into rationalizing a sales portfolio in a way that steers the strategic direction of the company. This is due, to a great extent, to the fear of a potential drop in short-term performance and the subsequent vulnerability to hostile takeover.

The blockbuster model is therefore under a number of threats to its long-term viability, as can be seen in Figure 10.3.

- Physician access becoming more difficult
- R&D moving to selective and targeted medicines
- Reconciling mass versus agility for the pharma organization difficult
- Unrationalized sales portfolios persist
- Big pharma, biotech, and generics houses are converging
- Strategic decisions take decades to pay off
- Complexity of communication is increasing

Strategic portfolio brand management
Development of a "pharmaceutical brand logic"

FIGURE 10.3. The blockbuster model is under attack.

BLURRING SEGMENTATION BETWEEN PHARMACEUTICALS, BIOTECH, AND GENERIC MODELS

The biotech segment is starting to look more and more like the pharmaceutical sector, an aspect that is shared with the generics segment of the industry, where development of risky NCEs is starting to happen. Figure 10.4 looks at the pharmaceutical industry's value curve and shows how the traditionally quite different segments (generic, patented, and biotech) are becoming more similar and their boundaries less distinct. Each segment participant eyes the profits that others make as the environment becomes more challenging—the one desire they all have in common is the wish to gain the most value, and that is generally with innovative new product brands.

Therefore, a general blurring of the boundaries across the industry segments is occurring, particularly when the business model of Novartis is taken into account. This company is a leading prescription branded company with its Novartis brand and, in addition, is allied to the second biggest generics house in the world through its Sandoz brand name.

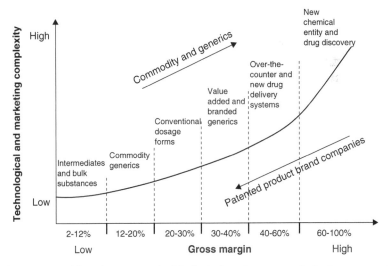

FIGURE 10.4. The pharmaceutical industry's value curve. Industry segments become blurred as everyone competes for the margins given by innovative brands.

Few biotech companies have transformed themselves into efficient sales and marketing organizations, the major exception being Amgen, which has managed to retain most of the value of its products through self commercialization. As a result, in 2003, it marketed five of the nine global biotech blockbusters. But this is rare, with the majority required to license out or sell their NCEs to generate enough cash to maintain their R&D activities. Having said that, consolidation is occurring within this sector as well, for example, the recent acquisition of the largest U.K. Biotech Company Celltech by UCB S.A. This Belgian hybrid then completed its transformation to be a pure biopharmaceutical player by selling its chemicals business (Surface Specialties) and combining the large-molecule R&D capabilities of Celltech with its in-house small-molecule expertise. The rationale being to grab all the value through commercializing the resultant molecules through its global sales and marketing network.

The generics sector has also seen a foray into risky NCEs over the last few years. This industry, whose innovation used to be constrained to new formulations of existing drugs, has moved through gaining experience with novel drug delivery systems and fixed combinations to NCEs. In a similar fashion to traditional pharmaceutical companies, some have achieved this via collaboration with biotech companies. Teva signed a 2004 agreement with Active Biotech to develop and commercialize a Phase 2 immunomodulator for relapsing multiple sclerosis. Others such as Apotex have already successfully commercialized their own first NCEs—Ferriprox (deferiprone) for treatment of iron overload in thalassemia patients—and see their own foray in the biotech arena as a preparation for the eventual regulatory acceptance of biogenerics in the future. Although the focus for these companies remains generics, an area of rapid growth in itself, they are using their R&D budgets (Teva, Hexal, Schwarz, and Barr spending above $150 million per year)[3] to update their business models as the market evolves.

STRATEGY DECISIONS TAKING DECADES TO PAY OFF

One of the key aspects of the pharmaceutical business model is the amount of time it takes for strategic decisions to pay off. The decision to focus on a particular therapy area or compound may take fifteen years to bear fruit, while in many other industries the payback can be measured sometimes in months and years, not decades. High-tech products in particular, such as mobile phones, can move from their conception in marketing to development in R&D and then onto the market in less than a year. In contrast, the decisions that made Lilly, the fourth-ranked seller of biotech brands in 2003, were made many years ago. It developed biotech capabilities early on and integrated the technologies effectively with its small-molecule-discovery efforts.

Another crucial factor is that of serendipity in R&D activities. It would have been difficult to predict that Pfizer's erectile dysfunction brand Viagra (sildenafil) was going to end up a lifestyle drug following its early development in the cardiovascular area. Serendipity is sometimes for the better, sometimes for the worse; therefore, it is a factor that is extremely difficult to predict.

A diagrammatical representation of the differences between pharmaceutical brands and FMCG brands is shown in Figure 10.5. It visually relates some of the differences seen in R&D, marketing preparation, and then the resultant product or brand created.

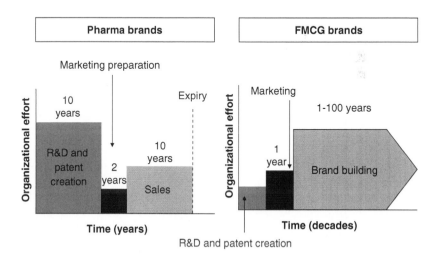

FIGURE 10.5. Diagrammatical representation of differences between pharmaceutical brands and FMCG brands.

COMPLEXITY OF BRAND COMMUNICATION INCREASING

In the global pharmaceutical market, brands need to act as communication tools between the pharmaceutical companies and their geographically separated customers, whether that is the patient, the physician or the payers. One of the phenomena of recent brand histories has been the creation of brands that are based on an attitude rather than a category. Amazon, Lastminute.com, and Yahoo! attract consumers to a Web site and are then able to sell them virtually anything. This illustrates the increasing complexity of communication in the modern world when one thing remains constant—satisfied customers become loyal customers and bring long-term profitability.

Will the blockbuster product model survive? The answer in the short term is an emphatic yes! But the writing is on the wall to significant change over the next three decades. The industry has, without doubt, got to find new avenues of wealth creation.

THE ANSWER?

Companies need to recognize that the winners of the future have to differentiate themselves and their products as unique for physicians, payers, patients (consumers), and investors. Single successes will no longer be sufficient for the big players, the size of the modern big pharma portfolio means that to sustain growth going forward, they have to launch new blockbusters every other year of so. At one extreme, Pfizer had approximately $48 billion of sales in 2003, growing by perhaps 10 percent means close to $5 billion in new sales needs to be added annually—a frightening amount of sales, considering that at the time there were only sixty-seven blockbusters across the industry. So far, the product has been treated "as everything," probably due to the vast amounts of R&D resources and corporate intellectual thinking that go into creating new cures and therapies for disease, and this is driven home to the smaller players who know just finding one blockbuster product can transform a company. Finding a curative drug, in an area of high unmet medical need, such as the treatment of chronic myeloid leukemia (CML), allows the innovator the luxury of free pricing decisions. Gleevec from Novartis was set at a worldwide price of $2,200 per month, and then won friends by introducing innovative pricing schemes and expanded access according to the ability to pay.

In general, investment is concentrated on creating the original product. Even worse, brand destruction is seen almost as a casual item for consideration. Pharmaceutical strategists are given the excuse for short-term thinking because of the classic patent expiry issue and a general lack of interest in changing the business model.

In those industries that have adopted a strategic branding approach (or brand logic), the "brand is everything" when major decisions need to be made. This is in stark contrast to what happens when it comes to setting the priorities and resource allocation in pharmaceuticals. Wealth is created by long-term brand investment, and not by a number of single products. In comparison, in fast moving consumer goods (FMCG) companies, R&D, while important, does not consume up to 20 percent of annual turnover, and the investment in sales force manpower is far lower too. Brand management maximizes the long-term value of the brand over decades. Pharmaceutical brand management needs to reach a new level of understanding and sophistication if it is going to create the positive impact that other industries have realized. A new model needs to be created.

CONCLUSIONS

So, what could that different model look like? Well, some simple things would be to formalize strategic brand management as a discipline within pharmaceutical companies. Global marketing at present is heading in the other direction by becoming closer to R&D and the pipeline. This may be a necessity of moving to a targeted future, but the management of the overall sales portfolio appears to be left to drift. (See Table 10.1)

Why allow brand destruction to occur? Sustaining a brand over time is achievable with the right thought and strategic thinking. It takes a mind-set change but is achievable if the right organizational direction is maintained.

We know that the brand communication processes are becoming more complex, and yet, as an industry, we haven't found the right mix for that communication. Be that representative with physician, DTC adverts with the patient, endorsement lobbying with the general public, or a whole host of otherwise unmanaged interactions that occur on a day-to-day basis.

TABLE 10.1. Pharmaceutical brand logic.

Formalize strategic brand management—think about brand architecture and how it could be implemented

Stop brand destruction

Focus on sustaining a brand over time

Challenge marketing to manage multiple customer interactions rather than just feed huge sales forces

Rationalize sales portfolios

Build brands: corporate, franchise, or product to improve long-term profitability

Prepare for the future—look at your cost base

Why not look again at the cost base that those mergers have created? In the first half of 2004, AstraZeneca spent more than twice its R&D budget on sales, marketing, and general administration—that doesn't appear to be the best investment for the future if we believe change is coming. Why not rationalize some of those vast sales portfolios? Adoption of somewhat basic portfolio branding guidelines would provide some obvious opportunities of synergies with the anticipated products coming through the pipeline.

Make marketing pay its way rather than being used as a department for feeding vast sales forces. Why not take some risks? Start calling products "product brands," and integrate them in a longer-term plan and portfolio of brands. The branded prescription industry could even reduce the price of their brands as generics approach and capitalize on years of experience and their rich patient knowledge to retain patients on the original brand, but cheaper.

Why not concentrate on building corporate brands for the top twenty players and generics houses, franchises for the mid-size players, and creation of product brands for the smaller companies and biotechs?

To some extent, the natural progression of the industry may well help the process—the evolution to targeted brands will change the status quo in how we sell and market pharmaceuticals. If you have a sufficiently specialist brand, the global target audience can be tiny and the cost of effective coverage will be within the reach of the smallest of biotech or even generics company. The days of merely expanding the number of salespeople to increase sales have ended, and, during the transition, the issue will be how best to maintain "the right size" while working on improving other means of communication— hopefully, the concept of branding, as a strategic imperative, will eventually get through.

Notes

Chapter 1

1. www.sandoz.com/site/en/company/heritage/content.shtml (January 2005).
2. Ellwood, I. *The Essential Brand Book* (London: Kogan Page, 2000), 58.
3. Moss, G. "Pharmaceutical brands: Do they really exist?" *International Journal of Medical Marketing,* 2 (2001), 1, 23-33.
4. Kapferer, J.N. "Marque et médicaments: le poids de la marque dans la prescription médicale," *Revue Française du Marketing* (1997), 165.
5. Chandler, J. "From product to brand in 40 years' (a short history of Western medicine)," personal communication (2004).
6. Chandler, J. and Owen, M. "Pharmaceuticals: The New Brand Arena." *International Journal of Market Research,* 44, Q4 (2002).
7. Datamonitor report on "Pharmaceutical Promotional Effectiveness," (November 2002).
8. Kapferer, J.N. "Marque et médicaments: le poids de la marque dans la prescription médicale," *Revue Française du Marketing* (1997), 165.
9. Chandler, J. and Owen, M. "Pharmaceuticals: The New Brand Arena." *International Journal of Market Research,* 44, Q4 (2002).

Chapter 2

1. Blumenthal, D. "For the end of brand balderdash—and the beginning of a real future," Editorial, *Journal of Brand Management,* 2 (2004), 3, 177-179.
2. Ibid.
3. Schuiling, I. and Moss, G. "A brand logic for Pharma? A possible strategy based on FMCG experience," *International Journal of Medical Marketing,* 4 (2003), 1.
4. Ellwood, I. *The Essential Brand Book* (London: Kogan Page, 2000), 17.
5. Kapferer, J.N. *Strategic Brand Management* (New York: The Free Press, 1991), 46. (Second edition, London and New York: Kogan Page, 1998).
6. Aaker, D.A. *Building Strong Brands* (New York :The Free Press, 1996), 7-8.
7. Ibid.
8. Chandler, J. and Owen, M. "Brand Innovation and factors of change." BHBIA conference (2002).
9. Aaker, D.A. and Joachimsthaler, E. *Brand Leadership* (New York: The Free Press, 2000), 8.

Pharmaceuticals—Where's the Brand Logic?
Published by The Haworth Press, Inc., 2007. All rights reserved.
doi:10.1300/5836_11

10. Kapferer, J.N. *Strategic Brand Management* (New York: The Free Press, 1991), 46.

Chapter 3

1. Kapferer, J.N. *Strategic Brand Management* (New York: The Free Press, 1991), 46.
2. Aaker, D.A. *Brand Portfolio Strategy—Creating Relevance, Differentiation, Energy, Leverage, and Clarity* (New York: Free Press, 2004), 13-14.
3. Ibid.
4. Kapferer, J.N.
5. Ibid.
6. Ibid.
7. www.bmsoncology.com/data/manonc.html, (2004).
8. Aaker, D.A. *Building Strong Brands* (New York: The Free Press, 1996), 243.
9. Cited in Scrip, PJB Publications Ltd "Several authorized Concerta generics in the wings?" S00867104 (January 14, 2005).
10. Aaker, D.A. *Brand Portfolio Strategy* (New York: The Free Press, 2004), 46.
11. Ibid.
12. Kapferer, J.N. *Strategic Brand Management* (New York: The Free Press, 1991), 326.
13. Cited in Scrip, PJB Publications Ltd "Sanofi-Aventis spins out Anti-infectives Company" S00866164 (December 2004).

Chapter 4

1. James R.G. "The bottom-line impact of corporate brand investment: An analytical perspective on the drivers of ROI of corporate brand communications," *Brand Management,* 8 (July, 2001), 6, 405.
2. *Business Week,* "The top 100 Brands." (August 2, 2004).
3. www.pfizer.com/are/investors_releases/2005pr/mn_2005_0119.cfm.
4. Wright, H. and Fill, C. "Corporate Images, Attributes and the UK Pharmaceutical Industry," *Corporate Reputation Review* 4 (2001), 2, 99-110.
5. Oral presentation by Steve Hawgood Regional Compliance Manager (North Asia) Eli Lilly at the Sixth Asia Pacific Pharmaceuticals Roundtable (Economist Conferences) Shanghai Nov (2004).
6. Cited in Scrip, PJB Publications Ltd "Pharma donates medicines to tsunami victims," S00868288 (January 4, 2005).
7. Blackett, T. "Why can't pharma be more like this? Few companies choose to manage their corporate brand, but with pressure mounting from various stakeholders, they need to start making a better impression," *Pharmafocus,* (July 2003).
8. Pisani, J. "Corporate Branding, Communications and Trust," oral presentation, Eyeforpharma Conference (March 22nd-23rd, 2004).
9. Houlton, S. "Sandoz lives again as a brand name," *Pharmaceutical Executive,* (April 2003), 17.
10. Cited in Scrip, PJB Publications Ltd "Wyeth backs plan to speed up fen-phen payouts" S00869012 (January 11, 2005).

Chapter 5

1. Aaker, D.A. *Brand Portfolio Strategy* (New York: The Free Press, 2004), 218-219.

2. Datamonitor report "Pharmaceutical branding strategies," 83 pp., October 2001.

Chapter 6

1. Kapferer, J.N. "Marque et médicaments: le poids de la marque dans la prescription médicale," *Revue Française du Marketing,* (1997), 165.

2. Levitt, T. "The globalisation of markets," *Harvard Business Review,* 61 (May-June 1983), 92-108.

3. Schuiling, I. "Think local, act local: is it time to slow down the accelerated move to global marketing?" *European Business Forum,* (Spring 2001), 5, 68-70.

4. Jain, S. "Standardisation of international marketing strategy: Some research hypothesis," *Journal of Marketing,* 53 (January, 1989), 70-79.

5. Wind, Y. "The myth of globalization," *Journal of Consumer Marketing,* 3 (Spring 1986), 23-26.

6. Douglas, S. and Wind, Y. "The myth of globalization," *Columbia Journal of World Business,* 22 (Winter 1987), 19-29.

7. Kapferer, J.N. *Strategic Brand Management* (New York: The Free Press, 1991), 206.

8. Terpstra, V. "The evolution of international marketing," *International Marketing Review* (Summer 1987), 47-59.

9. Kapferer, J.N. "Is there really no hope for local brands?" *The Journal of Brand Management,* 9 (2002), 3. 63-170.

10. Sexton, O. "Global brands may not improve product ROI, but investors demand them," Datamonitor report (2002).

11. Naomi Klein, *NO LOGO* (London: HarperCollins Publishers, 2000).

12. Holt, D.B., Quelch, J.A., and Taylor, E.A. "How global brands compete," *Harvard Business Review,* September 2004, 68-75.

13. Lilly, R. "A pfriendly pface—Does Euro-Pharma actually exist? Or has big pharma become too big for boundaries?" *Pharmaceutical Marketing Europe,* Autumn 2004, 18-19.

14. Lam, M.D. "A $20 Billion Bill and Plenty of Change," *Pharmaceutical Executive,* September 2004, 112.

15. Wyke, A. "Market analysis. Direct to consumer marketing. Patient Views: DTC advertising," *International Journal of Medical Marketing,* 4 (2004), 4, 310-312.

16. Mehta, A. and Purvis, S.C. "Consumer response to print prescription drug advertising," *Journal of Advertising Research,* 43(2003), 2, 1-15.

17. Interbrand annual rankings 2002.

Chapter 7

1. Aaker, D.A. *Managing Brand Equity* (New York: The Free Press, 1991), 76.

2. Committee on the value of advertising, American Association of Advertising Agencies, "The value side of productivity: A key to competitive survival in the 1990s," (1989), 18.

3. Kapferer, J.N. *Strategic Brand Management* (New York: The Free Press, 1991), 321-336.

4. Ebsworth, J. "Blockbusters blast a highway through pharma sales," *IMS World Review*, 2003, www.ims-global.com/insight/news_story/0203/.

5. Cited in Scrip PJB publications Ltd "US appeals court reinstates Rezulin suit," issue 2843, pp. 20, (April 23, 2003).

6. GSK press release June 7th, 2002, www.gsk.com/press_archive/press_06072002.

7. Cited in Scrip PJB publications Ltd "Major recall following breaches at Pan Pharmaceuticals," issue 2846, pp. 21, (April 30, 2003).

Chapter 8

1. Cited in Scrip PJB publications Ltd "Can products with staying power be identified?" S00839192, (April 19, 2004).

2. Mick Kolassa, "Pharmaceutical life cycle management," oral presentation, August 2003.

3. Cited in Scrip PJB publications Ltd 'Can products with staying power be identified?' S00839192, (April 19, 2004).

4. World Pharmaceutical Market Summary–Issue 1/2005, e-mail received from IMS Health (service@open.imshealth.com).

5. Mascha, F. "Key factors to make a pharmaceutical product a mega brand," personal communication (February, 2001).

6. Ebsworth, J. (2003) "Blockbusters blast a highway through pharma sales"IMS world review, www.ims-global.com/insight/news_story/0203/.

7. Aitken, M and Holt, F. (2000) "A prescription for direct drug marketing," McKinsey Web site: www.mckinseyquarterly.com.

8. Cited in Scrip PJB publications Ltd "UK to have first OTC statin" S00843442, (May 12, 2004).

9. Tufts Center for the Study of Drug Development (2004) "2004 Outlook," www.csdd.tufts.edu/InfoServices/Publications.

10. Visiongain publishing, London, report titled "The future of Drug Delivery" (May 2004).

11. www.tap.com/npr_2003_11_19.asp. (April 2004).

12. Erickson, D. "Branding goes global," *The Business and Medicine Report*, (2001), 60-71.

13. Ibid.

Chapter 9

1. "Drug patents under the spotlight: sharing knowledge about pharmaceutical patents" *Medecins Sans Frontieres*, (May 2003).

2. Burstall, M.L. and Reuben, B.G. "The Outlook for the World Pharmaceutical Industry to 2015," *Decisions Resources* (2004).

3. Cited in Scrip, PJB Publications Ltd "Bristol-Myers Squibb to pay $670 million over BuSpar, Taxol suits" S00784314 (January 9, 2003).

4. Reported in *Pharmafocus,* 5 (March 2003), 3.

5. Burstall, M.L. and Reuben, B.G. "The Outlook for the World Pharmaceutical Industry to 2015," *Decisions Resources* (2004).

6. Ibid.

7. Cited in Scrip, PJB Publications Ltd "Pharma takes on counterfeiters in China" S00846446, (June 8, 2004).

8. Cited in Scrip, PJB Publications Ltd "Latvian pharma deals with ban of marketing–update" S00844984, (May 25, 2004).

Chapter 10

1. Simon, F. and Kotler, P. *Building Global Biobrands—Taking Biotechnology to Market* (New York: The Free Press, 2003), 134.

2. Tufts Center for the Study of Drug Development (2004) "2004 Outlook," www.csdd.tufts.edu/InfoServices/Publications.

3. Buggle, I. "Foresight or folly? Generics firms' foray into NCEs," September, 30, 2004 www.open.imshealth.com/IMSinclude/i_article_20040930.asp.

Index

Page numbers followed by the letter "f" indicate figures; those followed by the letter "t" indicate tables.

Pharmaceuticals—Where's the Brand Logic?
Published by The Haworth Press, Inc., 2007. All rights reserved.
doi:10.1300/5836_12

Printed and bound by CPI Group (UK) Ltd, Croydon, CR0 4YY

17/10/2024

01775687-0007